THE ENERGY OF NATURE

# THE ENERGY

# OF NATURE

*E. C. Pielou*

THE UNIVERSITY OF CHICAGO PRESS • CHICAGO AND LONDON

E.C. PIELOU, former professor of mathematical ecology and Killam Professor at Dalhousie University, has been a naturalist all her life. She is the author of many books, most recently *Fresh Water*, *A Naturalist's Guide to the Arctic*, and *After the Ice Age*, all published by the University of Chicago Press.

The University of Chicago Press, Chicago 60637
The University of Chicago Press, Ltd., London
© 2001 by The University of Chicago
All rights reserved. Published 2001
Printed in the United States of America

10 09 08 07 06 05 04 03 02 01    1 2 3 4 5
ISBN: 0-226-66806-1  (cloth)

*Library of Congress Cataloging-in-Publication Data*
Pielou, E.C.
   The energy of nature / E. C. Pielou.
      p. cm.
   Includes index
   ISBN 0-226-66806-1 (alk. paper)
      1. Force and energy.  I. Title
   QC73.P54 2001
   530—dc21                          00-048841

∞ The paper used in this publication meets the minimum requirements of the American National Standard for Information Sciences—Permanence of Paper for Printed Library Materials, ANSI Z39.48-1992.

*In memory of Patrick*

# CONTENTS

When Robert Louis Stevenson wrote in a lighthearted vein, "Life is so full of a number of things / I'm sure we should all be as happy as kings," he mentioned only things and not events. This book is about events in the natural world—all kinds of events. They are as numerous and as interesting as things, and much more thought provoking. The salient point about events is that without energy they couldn't happen. Without energy, nothing would ever happen. Energy is as indispensable an ingredient of the universe as matter is. It is extraordinary that mentioning the word "energy" makes most people envision only power stations, hydroelectric dams, the price of oil, or athletes.

I consider energy from the point of view of a naturalist. To me "natural history" consists of more than the study of mammals, birds, butterflies, trees, and flowers plus thousands of other living organisms. The subject also includes the study of weather, of rivers and lakes, the oceans, the structure of the land, and much more: everything in which movement is visible or in which you know movement is happening although you can't see it. The movement may be too slow, as in tree growth and mountain building, or concealed, as in molten rock flowing deep underground, or invisible, as electric charges building up on clouds and on the earth's surface below.

The book contains no math apart from the occasional arithmetic calculation. The units in which speed, density, power, and the like are measured are written in scientific notation, as explained on page ix. It takes only a moment to grasp the principles, and any other style would be intolerably long-winded. The level of the book is about the same as that of articles in *Scientific American* or *New Scientist*.

As always, I am indebted to my husband, Patrick, and my editor at the University of Chicago Press, Susan Abrams, for contributing brainwaves and encouragement.

# SOME NOTES ON SCIENTIFIC NOTATION

*Powers of ten and of one-tenth*

Recall that $100 = 10^2$, $1{,}000 = 10^3$, and so on.

Likewise $0.1 = 1/10 = 10^{-1}$, $0.01 = 1/100 = 1/10^2 = 10^{-2}$, $0.001 = 1/1{,}000 = 1/10^3 = 10^{-3}$, and so on.

*Measurements*

Area: One square meter (m) is written as $1$ m$^2$.

Volume: One cubic meter is $1$ m$^3$.

Speed: One meter per second (s) is best written $1$ m s$^{-1}$ (not $1$ m/s).

Acceleration: One meter per second per second is best written $1$ m s$^{-2}$ (not $1$ m/s$^2$).

Density: One kilogram (kg) per cubic meter is best written $1$ kg m$^{-3}$ (not $1$ kg/m$^3$).

And so on, for any unit that would be spoken aloud as (something) per (something).

An exception to the rule: One kilometer per hour is $1$ km/h because neither the kilometer nor the hour belongs to the international system of units (ISU).

# 1 ENERGY IS EVERYWHERE

*In The Beginning*

Once upon a time, about 15 billion years ago, the universe—or more cautiously *this* universe—was brought into existence by the Big Bang. At the very first moment, it had zero volume and must have consisted entirely of radiant energy. The density of the energy would have been infinite, and the temperature was of the order of $10^{32}$°C. It immediately began to expand and to cool, and it has been doing so ever since.[1] As soon as its volume exceeded zero, things began to happen. By the time the infant universe was $10^{-43}$ seconds old (that's 0.00 . . . 001 with forty-four zeros), it had grown appreciably, but it was still smaller than a pinhead, about one millimeter in diameter. It was a tiny, expanding fireball, exceedingly dense and intensely hot. A very small fraction of its energy had become matter. From that day to this, energy plus matter has constituted the whole content of the universe.

While the universe aged from five minutes old to about 100,000 years old, it consisted almost entirely of radiant energy plus a plasma of hydrogen nuclei (protons), helium nuclei, and free electrons; no atoms existed until after the end of this stage. Keep in mind that the universe was 100,000 years old about 14,999,900,000 years ago; it had completed a mere 0.00007 of its life span to date and was still in its infancy. By the end of this stage the temperature had dropped to about 100,000°C, making it possible for protons and electron to combine in pairs and create hydrogen atoms. (That the 100,000-year-old universe had a temperature of 100,000°C is simply a coincidence; the numbers are only approximate in any case.) Thereafter galaxies and stars started to form, and the energy in matter began to exceed the energy in electromagnetic radiation. The changes have been continuing in the same direction ever since—less and less radiant  energy and more and more energy in matter—and will no doubt go on doing so. That is, conditions in the universe haven't changed qualitatively for the past 14,999,900,000 years; but its temperature continues to drop, it continues to expand, and matter becomes increasingly dominant relative to radiant energy.

After that brief account of the history of the cosmos, let us return to life on our planet. It is a world where things happen, and happenings always entail energy. Even the moon is not truly a dead world, in spite of its bad press. It may lack life in the usual sense of the word, but things happen there: meteorites strike it; the surface heats under the sunshine and cools during darkness, making the rocks alternately expand and contract so that they fracture; the fragments fall. And whenever anything is happening, energy is being transferred from one piece of matter to another.

It surely follows that energy should attract the attention of observers at least as strongly as "things" do. Everybody is surrounded all the time by energy transfers: events, actions, "happenings." It's worthwhile to consider the implications, especially for naturalists.

## A Hike in the Country

Imagine a hike in the country and the things an observant hiker would see. The list will probably include many living things: trees, flowers, birds, butterflies, perhaps squirrels and deer. There will also be scenery: rivers and streams, lakes, ponds and marshes, mountains and hills, perhaps beaches and the sea, and for skywatchers, blue sky and clouds by day or the moon, the stars, and maybe (with luck) a comet by night. The list can be extended almost indefinitely. It is a list of *things*, however—material things—and it represents no more than half of what surrounds the hiker. The scene is also filled with energy: not directly visible, it is true, but rendered observable through countless actions, movements, and events.

Imagine the scene once more, this time concentrating on all the signs of energy to be seen: twigs and branches swaying in the wind, scudding clouds, flowing water, breaking waves, flying birds and insects, running deer. Things both living and nonliving are continually moving, a sure sign that energy is being spent. Think of the sounds the hiker hears, for sound is a form of energy: the crackle of dry leaves underfoot or the drumming of rain, the babbling of a stream, the calls of birds, the hum of insects. Sound is much more noticeable on a windy day, with the roar of wind and waves at the beach and the snapping of tree branches in the forest. The stormier the weather, the more obvious the energy. Lightning gives a glimpse of yet another of energy's many forms—electrical energy.

Movements, sounds, and the occasional lightning flash are merely the more attention-getting forms of energy. The warmth and brightness of sunshine and the growth of plants illustrate how the sun's energy empowers life and action at the surface of the earth; energy from the sun comes as electromagnetic radiation, and plants grow because they can convert the radiant energy into chemical energy.

Energy in a multitude of forms is as much a part of our surroundings as are tangible things, and it is just as noticeable to anybody who pays attention. In the city, evidence of energy at work—man-made energy—is impossible to avoid: think of the roar of traffic, the bright lights, the construction sites with cranes and concrete mixers, even the din of shopping-mall music. But energy is as abundant in the tranquil countryside as it is in the city, since all energy has its ultimate origin in natural sources exactly as material substances do. Imagining otherwise is like a city child's not believing that milk comes from cows because it so obviously comes from cartons.

Energy is as much a part of nature as matter, and all artificial energy derives from natural energy. Coal, oil, and natural gas are stores of fossil solar energy. Hydroelectric power is simply solar energy that has been converted to human use more quickly. Nuclear energy existed as natural energy for billions of years before humans built nuclear power plants. Knowledge about energy is knowledge about the basic workings of the universe and is fundamental to all of science; it is not simply part of engineering. Name any branch of science—physics, chemistry, biology, geophysics, oceanography, meteorology, quantum mechanics—and you will find it is about energy as much as about matter. From black holes and supernovas to viruses and genes, "things" of all kinds have both energy and matter; their energy is as important a part of them as their matter.

We will now begin a systematic look at the various kinds of energy and how they act in the natural world.

# 2 WHAT IS ENERGY? SOME PRELIMINARY PHYSICS

*Some Definitions*

In answer to the question, What is energy? no less a scientist than the late Nobel laureate Richard Feynman said, "In physics today, we have no knowledge of what energy *is* . . . . It is an abstract thing."[1] That was in 1963. At a profound epistemological level it is no doubt true to this day. In the same philosophical vein, it is equally true of matter. But for practical purposes that answer is not much help.

Turning to more mundane sources, we find that energy is "the capacity . . . to perform work," which is hardly a stand-alone definition. To be complete, it requires a definition of *work*. From the same source, the definition of work is "energy transferred to or from a body . . . . it involves an applied force moving a certain distance."[2] This circularity is unavoidable: in simple terms, work requires the expenditure of energy, and energy spent performs work.

Let us look more closely at *work*, the application of a force through a distance. It helps to consider an actual example. To pick up a five-kilogram block of iron from the ground and raise it to a height of two meters is work: it requires energy. Force must be exerted—enough force to overcome the gravitational pull of the earth on the 5 kg block; the force must be applied directly upward, against the pull of gravity, for a distance of 2 m. We can measure this amount of work by multiplying the force times the distance through which it acts; the answer measures both the work done in lifting the block and the energy required to lift it: they are the same. It remains to consider how *force* is to be measured.

Force is what it takes to accelerate a mass. If your auto has run out of gas and you want to push it along a level road, it takes considerable force to get the movement started—to accelerate the auto from zero speed to walking speed—but hardly any force to make it continue rolling at walking speed; once it is moving steadily, no force is required beyond that necessary to overcome any slight roughness of the road and any friction in the bearings. If there were no roughness and no friction, the force needed to keep the auto moving forward at an unchanging speed would be zero.[3]

Now let's return to the 5 kg block being lifted from the ground: the force of gravity (the force you are working to overcome) imparts acceleration to anything it acts on, and at the surface of the earth this acceleration, known as gravitational acceleration,[4] is 9.81 meters per second per second (briefly, 9.81 m s$^{-2}$; see page ix for an explanation of the symbols). This means that if you drop an object from a height (as Galileo is said to have done from the Leaning Tower of Pisa), it will fall at an ever increasing speed. It is being accelerated by the force of gravity acting on it. If the object is heavy enough for air resistance to be negligible, it will be falling at a speed of 9.81 meters per second (9.81 m s$^{-1}$) after one second, twice that, or 19.62 m s$^{-1}$, after two seconds, 29.43 m s$^{-1}$ after three seconds, and so on; the speed keeps on increasing steadily. This is true whatever the mass of the object. A measure of the amount of force acting on it is given by multiplying the acceleration by the object's mass.[5] The answer is in *newtons* (abbreviated as N); one newton is the force required to give a mass of one kilogram an acceleration of 1 m s$^{-2}$.

Therefore, when you hold a 5 kg block you are exerting an upward force of $5 \times 9.81$ N = 49.05 N. If you stop exerting this force, the block falls to the ground.

An aside is necessary here, to explain the difference between *mass* and *weight*. At the surface of the earth, an object's mass and its weight are the same by definition. For example, a 50 kg woman has a mass of 50 kg, and she weighs 50 kg; to use both terms seems mystifying and redundant, or at least

WHAT IS ENERGY? 7

it did to schoolchildren in the days before space travel. However, if the woman travels to the moon her mass will not change—it will still be 50 kg—but she will weigh much less, specifically 8.5 kg. The 8.5 kg is the force, confusingly called "weight," that holds her to the moon's surface, where the acceleration due to gravity is only 1.67 m s$^{-2}$, which is 17 percent of the acceleration on earth.

Now let's return to the topic of *work*, specifically the work required to raise the 5 kg block vertically through 2 m. This is equivalent to exerting a force of 49.05 N through a distance of 2 m. The answer is force times distance, and the resultant energy, measured in *joules*, is 49.05 newtons $\times$ 2 m = 98.1 joules.

Joules are the units in which both *work* and *energy* are measured. Thus one joule is the work done when a force of one newton is applied over a distance of one meter. It is also the energy expended in doing the same thing. Joules will be used throughout this book as a measure of energy. The abbreviation for them is simply J. To compare the energies of, say, earthquakes, rising and falling tides, breaking waves, sunlight falling on a patch of ground, the sunlight trapped by photosynthesis needed to grow a tree, the sound of thunder— whatever it is—one needs a unit for measuring energy, and that unit is the joule.[6] It is not, admittedly, a unit familiar from frequent use in everyday life, as is true of kilograms (for measuring mass), meters (for measuring length or distance) and seconds (for measuring time). But once you concentrate your attention on *energy*, the unit soon becomes familiar: you get used to it.


*Energy Conversions*

Energy exists in many forms. Electrical energy, electromagnetic energy, chemical energy, heat energy, and nuclear energy are only a few. Moreover, any form of energy is convertible into any other, though not necessarily at a single step. Most of the actions going on in the world involve several energy conversions.

Here is an ecological example. The sun generates its energy by nuclear fusion, which yields enormous amounts of radiant energy (light, heat, and ultraviolet rays); this energy leaves the sun in all directions as *electromagnetic energy*, a small fraction of which strikes the earth. Suppose some of this solar energy falls on a tract of grassland. The grass uses the solar energy to create sugars by the process of photosynthesis. That is, the chlorophyll in the grass converts electromagnetic energy into *chemical energy*. The grass grows—entailing a whole series of conversions of chemical energy—until some of it is eaten by a jackrabbit.

The jackrabbit leads an active life; to acquire chemical energy to fuel its own activities, reproduction, and growth, it must eat. It must hop hither and thither, biting off blades of grass and chewing them. That is, its limbs and jaws move: chemical energy in the jackrabbit's muscles has been converted into *kinetic energy*, the energy of movement. Eventually a coyote catches and eats the jackrabbit; this requires a fairly lavish conversion of chemical energy into kinetic energy by the coyote, since the jackrabbit will no doubt resist. Both the animals are warm-blooded, and to keep their temperatures at the physiologically correct level, they must also convert some of their chemical energy into thermal energy. Death finally claims the top predator, the coyote; some of its remains are consumed by scavengers, and what's left decays—it is consumed by decay organisms, chiefly bacteria and fungi. These, though not warm-blooded, still produce heat as a by-product of their activities. In the end the solar energy that was first captured by the grass is finally dissipated as waste heat.

This short story, with many details glossed over—or it would have taken pages and pages—could also have been written as the life history of a joule. Instead of treating it as a tale about a series of different objects—sun, grass, jackrabbit, coyote, bacteria—we could have made it the tale of a single unit of energy, a joule, and the conversions it underwent in a sequence of different settings before ending up, as all energy eventually does, as heat. We return to this ultimate fate of all energy in chapter 3, under the heading *entropy*.

Change of any kind, anywhere, entails energy conversion of one sort or another. Whenever you see energy being spent in movement—in the flight of a bird, the breaking of a wave, or the flow of a river, for example, it is worth asking how and where the energy originated and how and where it will be dissipated.

## Potential Energy

Let's return to the 5 kg block. It was lifted from the ground and placed on a shelf 2 m up (unless you're still standing there holding it). Work was done on it—specifically, 98.1 J of work. It has been given energy, but in spite of that it stolidly sits there, motionless, on the shelf. Where has the energy gone? The answer is that it has become *potential energy*, or PE for short. If the shelf gives way, the block will fall back to the ground; that is, the PE you gave it by lifting it will be converted back to movement—kinetic energy.

The form of PE possessed by the 5 kg block is known as *gravitational PE*. Anything poised to fall if something gives way has it—a leaning tree, a boulder on a clifftop, the water behind a dam. But what if the leaning tree is strongly rooted or the boulder is in the middle of a flat plateau, so that neither can truly be called "poised" to fall? Their collapse is not imminent. Does this make a difference to their gravitational PE? Surely the energy is not a mere matter of chance.

No, it's not. Gravitational PE is a relative matter. If one chooses to treat the surface of the earth at mean sea level as the level at which gravitational PE is to be regarded as zero, then anything whatever above that level has measurable PE, whether or not it's poised to fall.[7] A person living on a plateau high above sea level might prefer to treat the plateau as the level at which gravitational PE is to be regarded as zero. Then a 5 kg block on a shelf 2 m above the floor in a house on the plateau would have the same gravitational PE as an identical block 2 m above the floor in a house with its floor at sea level.[8] But if one chose to use sea level as the reference level for measuring the gravitational PE of both blocks, and if the elevation of the plateau is, say, 250 m, then for the block in the house at the seaside, the gravitational PE would be 98.1 J as before, whereas the PE of the block in the house on the plateau, on its shelf 252 m above sea level, would be

$$49.05 \text{ N} \times 252 \text{ m} = 12,360.6 \text{ J}.$$

Likewise, a rock below sea level, in Death Valley, say, has negative PE relative to sea level; energy would have to be spent to raise it to sea level.

This demonstrates that measurements of potential energy are arbitrary. The reference level against which gravitational PE is measured is always a matter of choice and must be stated if there could be any doubt.

Energy is stored as PE in a multitude of ways. A stretched spring or an archer's drawn bow stores *elastic energy:* the stretched spring snaps back to its unstretched length when let go; a stretched bowstring straightens when released, speeding an arrow on its way. In both cases, stored elastic energy has changed to kinetic energy.

Another familiar form of potential energy is *chemical PE.* An electric battery and a loaf of bread both have it. The conversion from potential to actual produces an electric current in the case of the battery and muscle movement in the case of the bread.

*Magnetic PE* is stored in magnets, ready to be converted to kinetic energy when a piece of steel is attracted to the magnet.

The list goes on: potential energy in its various manifestations will appear frequently in all that follows.

*The Ideal and the Real*

In theory (though never in practice), certain actions go on forever. Here are two examples; in both, gravitational PE is equal to zero at the lowest level reached by the moving object.

First, imagine a pendulum suspended from a perfectly frictionless bearing swinging from left to right and back again (fig. 2.1). Its bob (the hanging weight) is suspended by a perfectly *in*elastic string. Assume that the pendulum has been set up in a perfect vacuum, so that its movements are not affected by air resistance. The pendulum will continue to swing forever without any loss of amplitude. It is intuitively clear that this should happen, even though the conditions prescribed for the experiment are too perfect ever to be attained in practice. What happens to the imaginary pendulum is this: when the bob is at the left extremity of its swing, it is motionless for an instant; that is, it has no kinetic energy (KE). All its energy is potential; more precisely, it is gravitational PE. Then the bob starts to fall because of the force of gravity, but it is constrained by the string to swing to the right; as it swings, its PE is converted to KE. By the time the bob reaches the bottom of its swing, its PE is zero, having all been converted to KE; at this instant its KE, and therefore its speed, has reached a maximum. Nothing stops the bob's continued movement, so it keeps on swinging to the right and begins to ascend, losing KE and gaining PE in the process. The conversion of KE back into PE continues as the bob approaches the right-hand end of its swing. Here the conversion is complete: the bob's KE has decreased to zero so that it is momentarily stationary, and its gravitational PE has increased to a maximum. Then the whole process happens again, from right to left. The total energy remains the same all the time, never dwindling; it is the sum of the KE and the PE, known as the *mechanical energy* of the pendulum. As an equation,

mechanical energy = potential energy + kinetic energy.

In the ideal case, the mechanical energy remains unchanged forever, and the pendulum keeps on swinging.

In real life, with conditions unavoidably less than perfect, this does *not* happen. Because of friction in the bearings, air resistance, and minute stretching of the string, energy is gradually drained away from the pendulum in the form of imperceptibly slight heating. The mechanical energy slowly declines, and the amplitude of the swings diminishes, until all movement stops. At this stage the pendulum's mechanical energy has all been dissipated and it hangs motionless.

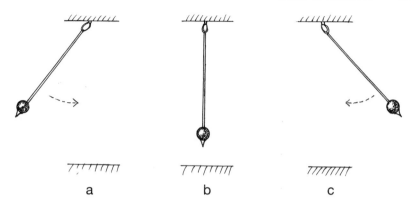

Figure 2.1. Three positions of a perfect (frictionless) pendulum. (a and c) Here the bob has maximum gravitational PE and zero KE. (b) Here it has maximum KE and zero PE.

For a second theoretical example, imagine a perfectly elastic rubber ball bouncing on a perfectly rigid floor in a perfect vacuum (fig. 2.2). The ball will continue to bounce forever, returning to the same height above the floor at each bounce.[9] As with the pendulum, the bouncing ball retains its total mechanical energy, which at every instant is the sum of its PE and its KE. The bouncing ball is slightly more complicated, however. Its PE is gravitational when it is at the top of its bounce and descending floorward and elastic when it recoils from the floor and starts upward. The KE of the ball is at a maximum on its downward journey just as it hits the floor. There the ball is abruptly stopped by the collision with the floor, but its KE is instantly converted to elastic PE and as instantly released, restoring the ball's KE, in an upward direction this time. The renewed KE and the upward speed of the ball are at a maximum just as the ball leaves the floor; they decrease to zero as the ball reaches its highest point.

In the real-life equivalent of this experiment, with an imperfectly elastic ball, an imperfectly rigid floor, and an imperfect vacuum (or none at all), we know that the bounces will steadily become lower and lower until they peter out altogether. That is, the ball's mechanical energy will be dissipated as heat, some of it in the air because of air resistance, and some of it in warming the imperfectly elastic ball and the imperfectly rigid floor; as these compress and expand, shearing within them causes friction.

The foregoing paragraphs have shown, implicitly, that energy results from

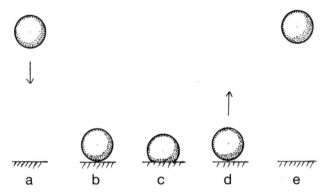

Figure 2.2. Five positions of a perfectly elastic bouncing ball. (a and e) Here, at the highest level reached at each bounce, the ball has maximum gravitational PE, zero elastic PE, and zero KE. (b and d) Here it has maximum KE (downward and upward respectively) and zero PE (both gravitational and elastic). (c) Here, where the ball is slightly flattened against the rigid floor, it has maximum elastic PE, zero gravitational PE, and zero KE.

two kinds of forces. One kind, exemplified by gravity and elasticity, is called a *conservative force;* its salient feature is that it can be stored—in these examples, as gravitational PE and elastic PE. A system in which the only forces acting are conservative forces never runs down. The other kind of force, exemplified by friction and air resistance, is *nonconservative.* When nonconservative forces are operating, either alone or in combination with conservative ones, a system inevitably runs down. Nonconservative forces produce heat, and the heat can never spontaneously turn back into another kind of energy.[10]

# 3 ENERGY AND ITS ULTIMATE FATE

## Friction and Drag

Friction is regarded with disfavor by most people except, possibly, the manufacturers of lubricants. Whenever something sticks that should slide, friction is to blame. Friction is an indispensable force, however: without it you could not walk or write; you could not make an auto move forward—the clutch would never stop slipping—and if you could the brakes would fail completely. Bedclothes would slide off the moment you got into bed. Friction's services are virtually endless, and they are all taken for granted.

Friction is equally indispensable in the natural world; a walk in the country provides unlimited examples. Take birds' nests: most are held together by friction and would fall apart without it. It is friction that allows a bear to flip a salmon from a stream, a cormorant to alight on a sloping rock, and a bighorn sheep to clamber over steep terrain. Again the list is endless.

One other force is as important as friction in impeding motion: it is *drag,* or more precisely *viscous drag.* The term in-

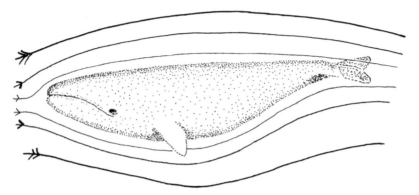

Figure 3.1. The no-slip phenomenon. The streamlines show how water flows past a swim-ming whale. The thickness of the lines represents the speed of the water relative to the whale; the flow speeds up at increasing distances from the whale, from zero at its skin (not to scale).

cludes both air resistance and water resistance, and it slows everything that moves through air and water. Drag is often treated as a form of friction, but the two are fundamentally different. Consider a whale swimming through the water: the water does not *slip* past the whale's flanks, in spite of appearances. At the surface where the skin of the whale and the water make contact, they stick firmly together because of a strong attraction between the molecules of a solid and a liquid. This surprising effect is known as the *no-slip* condition.[1] It prompts the question, *How* does a whale move effortlessly through water if slippage does not take place?

The answer is that all the slippage takes place by shearing movements within exceedingly thin layers of water encasing the whale (see fig. 3.1). This *viscous shearing* brings about a progressive increase in the velocity of the water relative to the whale, from zero right at the interface between whale and water. Molecules of water stick to each other and resist the shearing to some extent—hence drag. But water sticks to a solid more tenaciously than it sticks to itself; this accounts for the no-slip phenomenon and for the fact that drag rather than true friction (the resistance to sliding between two solids) impedes the motion between solids and fluids. The no-slip condition applies to every motion between a fluid and a solid: for example, the flow of a river over its bed, the flow of the wind past a crag, and the flow of air past a flying bird. It also applies to relative motion between a liquid and a gas, for example, a raindrop falling through the air.

The property friction and drag have in common is that both are nonconservative forces (see the final paragraph of chapter 2); that is, they cannot be stored as potential energy of one kind or another for later retrieval. Rather, they "go to waste" and produce "useless" heat. These common phrases conceal some fundamental facts of physics, as we shall see in what follows.

*Heat and Work*

Unlike a force such as gravity, which causes an object to accelerate, friction does the opposite: it resists motion and thus generates heat. The classic example of generating heat by friction is starting a fire by spinning a hardwood stick with its pointed end pressed against a wooden block. A clearer manifestation of mechanical energy being turned into heat would be hard to find.

In the 1840s the great British physicist James Prescott Joule carried out experiments to find how much work has to be done to produce a given amount of heat. At that time heat was measured in *calories;* one calorie is the amount of heat needed to raise the temperature of one milliliter of water by one degree Celsius. More precisely, it is the amount of heat needed to raise the temperature of 1 ml of water from 14.5°C to 15.5°C; this allows for the fact that the heat required varies slightly depending on the starting temperature.

Joule's most famous experiment was remarkably straightforward. He arranged for paddles submerged in a container of water to rotate under the action of a falling weight; knowing the weight, and the distance it fell, he could calculate the work done by the moving paddles. The viscous drag of the paddles stirring the water caused it to heat up, and the change in temperature was recorded. It was then possible to state how much work produced how much heat. Refined modern measurements show that, using joules (J) as the unit for work, 1 calorie (cal) is produced by 4.186 J of work. This is known as the *mechanical equivalent of heat.*[2] Equivalently, 1 J = 0.2389 cal.

The calorie is not yet obsolete as an energy unit, as any dieter knows. The unit listed as a Calorie (with a capital C) on food packages is equal to a thousand calories, that is, one kilocalorie. The energy in a slice of whole-wheat bread, for instance, is said to be 71 Cal, or about 300,000 J; this does *not* mean that eating a slice will fuel that much useful work—the efficiency of conversion must be taken into account.

*Heat and Temperature*

Imagine something—anything—gaining heat; it could be the air in a room when the sun shines in, the water in a kettle put on a hot stove, or a wooden block heated by friction when rubbed briskly with a stick. In every case, a gain of heat brings a rise in temperature of the object heated. This is so obvious that it seldom calls for comment. The problem that arises in a scientific mind, however, is this: What has changed within the heated object to make its temperature greater? The question is meaningless until we attach a meaning to *temperature*.

As nearly everybody knows, the molecules any object consists of, be it a roomful of air, a kettleful of water, a wooden block, or anything else, are in constant random motion. They are never still. And when an object's temperature rises, all its molecules speed up.

For a gas, the link between its temperature and the velocities of its molecules is surprisingly straightforward. The temperature of a mass of gas depends wholly on the average *kinetic energy* of its molecules.[3]

The kinetic energy (KE) of a body of mass $m$ moving with velocity $v$ is given by the formula $KE = \frac{1}{2} mv^2$. All the molecules in a gas collide with each other repeatedly; at each collision, the two colliding molecules bounce off each other in new directions and at altered speeds. Consequently each molecule's kinetic energy changes every time it collides, but this doesn't affect the rule, which relates to the average KE of *all* the molecules. Furthermore, the average KE (averaged over time) is the same for every molecule whatever its weight; therefore lightweight molecules must move (on average) faster than heavy ones. For example, if a given molecule is one-fourth as heavy as another, its average velocity must be twice as great.

Another way of wording the rule is to say, "When we measure the temperature of a gas, we are measuring the average . . . kinetic energy of its molecules."[4] This is the meaning of the word "temperature." It also makes clear what the *absolute zero* of temperature is. It is the temperature at which all the molecules have zero energy because they are motionless. This does *not* imply that subatomic particles are also motionless. Even at absolute zero, which is -273°C, they continue to oscillate, perpetually.

Absolute zero is used as the zero of the *absolute temperature* scale, also called the *thermodynamic temperature* scale; each division of the scale, a *kelvin* (abbreviated as K), is of the same magnitude as a degree Celsius. Thus the temperature of freezing water (0°C) is 273 K, and that of boiling water

(100°C) is 373 K. Blood temperature in a healthy human (37°C) is 273 + 37 = 310 K, and so on.[5] Note that the units are called "kelvins," not "degrees Kelvin."

That the molecules of a gas at a temperature above 0 K are forever randomly moving, colliding, and changing direction makes one wonder: What is their *mean free path?* In other words, how far does a molecule travel, on average, between one collision and the next? The answer depends on the amount of space available to the molecules, which depends in turn on the density of the gas. It is not dependent on the temperature of the gas, or equivalently, on the molecules' velocities: slow-moving molecules will take longer to travel from one collision to the next, but the distance traversed is the same. The density of the air is greatest at sea level and decreases rapidly at higher and higher altitudes.[6] At sea level, the mean free path of a molecule of air is 0.1 μm (micrometer); at an altitude of 100 km above sea level, it is 0.16 m; and at an altitude of 300 km, it is 20 km. This means that the mean free path of an air molecule 300 km up is $2 \times 10^{11}$ times greater than at the surface.

At an altitude of 300 km, in the farthest fringes of the atmosphere, the air temperature is about 1,500°C. How would this feel if we could experience it? It would unquestionably seem bitterly cold, because the familiar relationship between true, measured temperature and the subjective sensation of temperature holds only if the density (or pressure) of the air is what we are accustomed to. Recall that the temperature of the air is a measure of the average kinetic energy of *each molecule,* regardless of the number of molecules in a given volume. Now imagine two parcels of air, one at ground level and the other 300 km up, and suppose they are at the same temperature. It is easy to see that the total energy in the parcel at ground level far exceeds the total energy in the high-altitude parcel, because the former contains so many more molecules than the latter; it is the *total* energy, not the energy *per molecule,* that determines how the air "feels," either warm or cold. Thus the statement that at an altitude of 300 km the air temperature is 1,500°C, while true, gives no idea of how the air at that height feels: temperatures in air at unfamiliar densities cannot be imagined because we have never had the opportunity to become accustomed to them.

## Heat and Internal Energy

As we have noted already, unless an object is at a temperature of 0 K, all its molecules are in constant random motion; the object has *internal energy.* This is not to say that heat and internal energy are the same thing, however. They are not.

The distinction between them is best appreciated by considering what happens when you heat a kettle of water on a hot plate. Heat passes from the hot plate into the water and increases the water's internal energy. But that is not all it does; in addition, the heat boils the water, and the steam produced rattles the kettle's lid—the heat has done work on the lid. To repeat: the added heat has done more than merely raise the water's temperature; it has also done mechanical work. This can be written concisely as an equation:

$$Q = U + W.$$

Here $Q$ represents the added heat, $U$ the increase in internal energy of the water, and $W$ the work done.[7]

We can learn more by considering the classic piston engine driven by steam. In outline, the piston engine of an old steam locomotive works like this: water is heated in a boiler with a coal fire under it, producing steam under pressure; the steam expands into hollow cylinders, forcing out sliding pistons within the cylinders. The movement of the pistons is converted by camshafts and linkages into the rotary motion of the wheels. Indirectly, therefore, heat from the burning coal turns the locomotive wheels. Only a fraction of the heat supplied is turned into mechanical energy, however; most of the rest is lost in the steam escaping into the atmosphere; note that the heat put in from the firebox is at a much higher temperature than the comparatively cool steam discharged from the funnel. All the same, the cool steam carries away a proportion of the heat supplied by the firebox.

This is the crux of the matter: high-temperature heat yields a mixture of mechanical energy and low-temperature heat; the latter is wasted energy, but waste cannot be avoided. The *thermal efficiency* of any heat engine is defined as work done ÷ heat absorbed. Both parts of the fraction are measured in joules. Thermal efficiency is always less than one.

The maximum efficiency theoretically possible is given by the formula

$$(T_H - T_C)/T_H,$$

where $T_H$ and $T_C$ are, respectively, the temperatures of the hot (input) steam and the cool (output) steam, measured in kelvins.[8] It is easy to see that this fraction could reach one only if $T_C$ were absolute zero, an unattainably low temperature.

Thus no engine can be 100 percent efficient. Note that this is *not* a consequence of friction; even if friction could be reduced to zero (impossible in practice), the maximum efficiency of any heat engine would still be less than one, because of the very nature of thermal energy, expressed in the famous second law of thermodynamics. According to the law, "It is not possible to change heat completely into work, with no other change taking place."[9] In brief, there are no perfect engines. Yet another way of expressing it is to say that the random motion of the molecules in a hot substance can never change, completely and spontaneously, into ordered, macroscopic motion.[10] The wasted heat that cannot be made to change into mechanical energy and do work is the form of energy known as *entropy*. The meaning of this famous term has been the topic of whole books. Here we can give it only a section.

## Entropy

The law of the conservation of energy tells us that energy can be neither created nor destroyed. As we have emphasized repeatedly up to this point, energy put into a system is always, without exception, passed on in the same or another form: it never disappears.

At the same time, it is *never* true that all the energy supplied to a system can be made to do useful work. Some is always dissipated as unavailable heat, at too low a temperature to serve as the heat source for a heat engine. This heat is *entropy*; it could also be called *useless energy*.

## The Impossibility of Perpetual Motion

We have now come across two entirely different obstacles to so-called perpetual motion. First, recall the swinging pendulum and the bouncing ball described at the end of chapter 2; if their energy came *only* from conservative forces, their motion would be perpetual. But in real life, nonconservative forces—friction and drag—are always acting as well, and the motions of the two devices are inevitably brought to a stop. The energy they lose in slowing down is converted into "useless" heat, that is, entropy.

Second, as we have seen, some of the energy produced by heat engines is always useless heat (entropy again). This follows from the fact that a heat engine cannot, by the second law of thermodynamics, ever be 100 percent efficient.

These two points lead to the inescapable conclusion that *although the total energy of the universe remains forever the same, the fraction of it that is entropy forever increases.*

Another way of saying the same thing is, first, to contrast "ordered energy," such as the kinetic energy of a moving macroscopic object, with "disordered energy," namely thermal energy, the disordered, random motion of individual molecules. The law just given then becomes: Ordered energy is always ultimately transformed, spontaneously, into disordered energy.[11] The converse is not true—disordered (thermal) energy is never spontaneously transformed entirely into ordered energy.

To put the matter in a nutshell, the universe is running down. Everybody ought to know this nowadays, but we are still sometimes exhorted to "conserve energy," as if we could do anything else. What needs to be conserved, of course, is potential energy, especially that stored as chemical energy in fossil fuels. Entropy does not need conserving; it is increasing all too fast. Governments should be urging us to conserve fuels and slow down, as much as possible, the transformation of their energy into entropy.

*Mythical Perpetual Motion Machines*

As we have just seen, two separate facts make perpetual motion machines impossible. Therefore the two supposed forms of perpetual motion machines are both nonexistent.[12]

The first kind of mythical machine is exemplified by a water wheel that powers itself. The water in one of the buckets that has reached the top of the wheel tips its contents into an empty bucket below it, driving the wheel onward unceasingly. This device must have been independently invented by generations of mechanically minded children. But it can never work in practice because rotation of the wheel is resisted by friction, and it could keep on rotating only by creating new energy—which from the law of the conservation of energy is impossible.

The second kind of mythical perpetual motion machine is a heat engine working with 100 percent efficiency. This is impossible because it would entail the complete conversion of heat into work, violating the second law of thermodynamics.

Accepting the inevitable—that all energy will ultimately be converted into entropy—it is time to consider what is happening, and will continue to happen for a very long time, here on earth. The earth is continuously supplied with external energy from the sun, and it also generates internal energy of its own. These are the topics to be considered in the rest of this book.

# 4 SOLAR ENERGY AND THE UPPER ATMOSPHERE

*Power from the Sun*

In comparison with some of the far larger stars to be seen on a clear, dark night, our sun is often airily dismissed as a second-rate star. All the same, its energy output is impressive; it produces $3.8 \times 10^{26}$ J (joules) per second, without interruption.

The rate at which a source yields energy is its *power*. Power is measured in watts (W), and one watt is one joule per second. Writing this as an equation, $1\,\mathrm{W} = 1\,\mathrm{J}\,\mathrm{s}^{-1}$. The sun's power therefore is $3.8 \times 10^{26}$ watts, a quantity known as the *solar constant*.[1]

The sun radiates in all directions, and only a tiny fraction of its output is intercepted by the earth, 150 million kilometers away. On average, the solar power received by the earth is 340 watts per square meter of surface[2] or, more concisely, $340\,\mathrm{W}\,\mathrm{m}^{-2}$. It is important to be aware of what the averaging entails. First, the averaging is over the whole surface of the earth: it allows for the difference between the polar regions where the sun never rises high in the sky and the tropics where the midday sun is not far from the zenith on every day of the year. Second, the 340 W m$^{-2}$ is averaged over time: over day and night, and also over all the days of the year. Averaging over the year has nothing to do with the weather (the incoming radiation is measured above the atmosphere) or the seasons (averaging over the whole of the earth's surface takes care of that). Rather, it allows for the earth's elliptical orbit, which brings it nearest to the sun in January and takes it farthest away in July; this causes the energy received by the whole earth to be above average in January and below average in July.

*The Solar Energy Budget*

Now consider the fate of this incoming energy. The first point to notice is that solar energy does not accumulate appreciably. The earth's *net* gain of solar energy over the year is close to zero, and were it not for global warming it would remain at zero, on average. If we take the long-term view, disregarding slight temporary climatic wanderings caused by atmospheric changes, it is safe to say that all the energy that comes in must go out. Over the past several hundred million years a certain amount of solar energy has, admittedly, become stored as fossil fuels. The amount is negligible, however; it has been estimated that the heat content of all known fossil fuel reserves represents no more than the solar energy intercepted by the earth in ten days.[3]

The way the incoming and outgoing energies balance each other is shown in figure 4.1. The incoming sunlight, shown in the left panel, is chiefly short wave radiation in the visible and near ultraviolet parts of the spectrum. On average, 30 percent of it is reflected back to space by clouds and does not contribute any heat to the earth. Of the remaining 70 percent (about 240 W m$^{-2}$), 19 percent is absorbed by the atmosphere, chiefly by the water vapor in it, and the remaining 51 percent by land and ocean combined.[4] The right panel shows what subsequently becomes of this 70 percent; it is radiated back into space again, as infrared radiation for the most part; some is reflected back as light.[5] Of the 51 percent absorbed and then reradiated by land and sea, 45 percent is absorbed again on the outward journey, this time by the atmosphere, where it is held temporarily, adding itself to the 19 percent of solar energy absorbed on the incoming journey. The atmospheric ingredients responsible for the absorption are the "greenhouse gases," primarily water vapor, carbon dioxide, and methane. The total energy reradiated by the atmosphere therefore becomes 64 percent of the original input. The remaining 6 percent still "owing" radiates as infrared rays, directly from the ground to outer space.

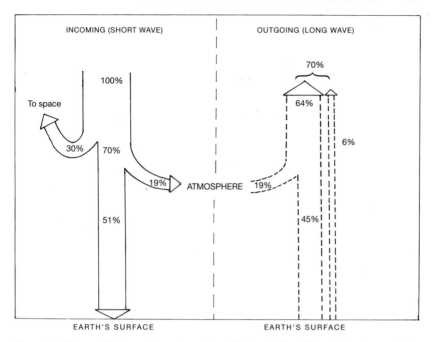

Figure 4.1. The partitioning of incoming solar radiation (on the left) and outgoing radiation from the earth (on the right). The latter adds up to the 70 percent of incoming radiation that was not reflected back to space.

Greenhouse gases are always naturally present in the atmosphere; if they were not, a much smaller fraction of incoming solar energy would be trapped to warm the earth, and a much larger fraction would be reflected directly back to space. If there were no atmosphere the earth's average surface temperature would be −18°C, that is, 33°C lower than the actual average of 15°C.[6]

Greenhouse gases in what humanity now thinks of as "natural" quantities—the quantities present before the Industrial Revolution—are an undoubted blessing; they are indispensable to our comfort, indeed, to our very survival. The global warming currently in progress is probably (not certainly) being brought about by the recent "unnatural" increases in greenhouse gases caused by pollution of the air with vehicular exhausts and effluent gases from a wide range of industries.

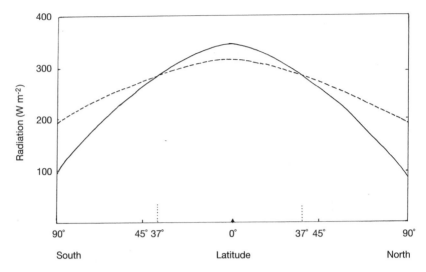

Figure 4.2. How incoming radiation (solid line) and outgoing radiation (dashed line) change with latitude. Between latitudes 37° N and 37° S, incoming radiation exceeds outgoing radiation. Poleward of these latitudes, incoming radiation falls short of outgoing.

*Latitudinal Temperature Differences and the Winds*

Solar radiation is what energizes the wind and controls the weather. The two most important factors governing the atmospheric circulation are the way air temperature varies from the equator to the poles and the way the earth's rotation on its axis affects wind direction. We'll consider these factors in turn.

The way the solar radiation reaching the earth's surface decreases as you go from the equator to the poles is shown by the solid line in figure 4.2. The incoming power ranges from a high of about 350 W m⁻² at the equator to a low of about 100 W m⁻² at the poles, for an average over all latitudes close to 240 W m⁻². The power decreases as the latitude increases because the angle of incidence of the sun's rays changes; in the tropics, the rays strike the ground almost perpendicularly much of the time, whereas at high latitudes they are always oblique.[7]

The dashed line on the figure shows how the absorbed energy is radiated back to space; it shows that absorption exceeds reradiation at latitudes between 37° N and 37° S and falls short of reradiation everywhere poleward of these latitudes. It follows that if it were not for redistribution of the sun's heat

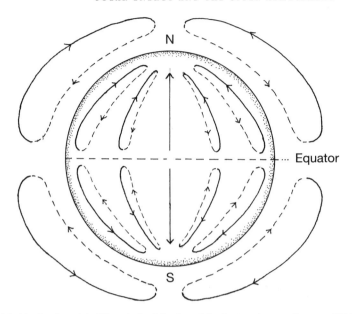

Figure 4.3. Idealized model of the winds at the top of the troposphere as they would blow if the earth did not rotate. The air would move poleward at high elevation (solid lines) and back toward the equator near sea level (dashed lines), circulating independently in the Northern and Southern Hemispheres.

by winds and ocean currents, the earth's climate would be entirely different: the tropics would be far hotter, and the polar regions far colder. In practice, however, air movements and ocean currents carry the sun's heat poleward from the tropics, reducing the climate contrast between high latitudes and low.

The general circulation of the atmosphere is controlled by the latitudinal temperature gradient, that is, by the way the temperature drops as you travel from low, tropical latitudes to high, polar latitudes. Here we consider the winds high in the atmosphere, far above the influence of friction with the surface (strictly speaking drag, but it is usually called friction). Air pressure depends on air temperature, being high where the temperature is high and low where the temperature is low.[8] Therefore the atmosphere develops a pressure gradient more or less matching the temperature gradient, with high pressures in the tropics and low pressures in the polar regions. Moreover, the greater the height above the earth's surface, the stronger the pressure gradient. The wind blows down a pressure gradient, from high pressures toward low. Consequently, if it were not for the rotation of the earth on its axis, high-level winds would tend to blow always from the equator toward the poles (see fig. 4.3); at the same time, to prevent the atmosphere from piling up over the poles, winds at the surface would blow from the poles to the equator, returning the air to its starting point. In other words, huge convection cells would develop, one over each hemisphere.

The worldwide pattern of circulation just described, and shown in the figure, represents what would happen, in theory, *if* the earth did not rotate on its axis once every twenty-four hours. But of course it does rotate, and the effect of this rotation is what we consider next.

### The Effect of the Earth's Rotation

Because of the earth's rotation any wind is deflected from its course unless it happens to be blowing parallel with the equator and directly above it. In the Northern Hemisphere the wind is always deflected to the right (as you stand with your back to it), and in the Southern Hemisphere, always to the left. This rule applies whatever the wind's direction. The deflection is known as the *Coriolis effect*.[9] The magnitude of the effect depends on the latitude: it is greatest at the poles and decreases to zero at the equator.

It is easy to see why the earth's rotation should cause a wind blowing over one of the poles, say the North Pole, to be deflected. Imagine yourself in a stationary satellite looking directly down on the North Pole; you would see the earth and everything fixed to its surface rotating counterclockwise beneath you. Suppose a weather balloon, floating high above the ground, was carried past on the wind directly below. The balloon is not attached to the earth and therefore does not move with it; instead, it is left behind by the continents and oceans carried along on the earth's surface so that, relative to them, it appears to drift westward, that is, to the right. The balloon would be seen to be going in a straight line if the earth below it were invisible; the rightward deflection is entirely a relative matter, relative to the earth and to an observer on the earth.

In this particular case, of an object moving southward from the North Pole, it is obvious how the Coriolis effect works. But it is not intuitively obvious how the effect can cause a free-floating object borne on the wind—and the wind itself—to be deflected to the right everywhere north of the equator, whatever the wind's direction and wherever the object may be.

A full explanation requires some fairly advanced mathematics, but figure 4.4 gives an idea of what is going on. It shows the globe rotating, once each day, around its axis (the line through its center joining the North and South

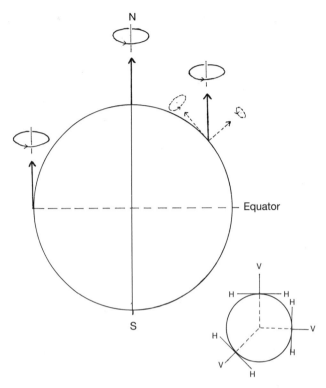

Figure 4.4. The Coriolis effect (see text). The earth is shown with three axes of rotation (solid arrows), the "true" axis and two "private" axes (see text). The perspective circles, with arrows, show the earth's spin at the three locations. The spin is wholly horizontal at the poles and wholly vertical at the equator. At intermediate latitudes it can be analyzed into horizontal and vertical components, as shown by the dashed arrows and circles at the midlatitude location. Inset: Horizontal (H–H) and vertical directions (V) at three representative points on the earth's surface.

Poles). If you were to stand at either of the poles for twenty-four hours in summer (when the sun never sets), the sun would appear to move in a complete circle around you, with the center of the circle directly overhead. Likewise, observed from any other point on the globe, the sun is seen to move in a complete circle in twenty-four hours, but the circle's center is *not* directly overhead. Admittedly, the sun is out of sight at night for an observer in non-Arctic latitudes, but visualizing where it would appear if the globe were transparent isn't difficult. It follows that every point on the globe can be thought of as rotating about a "private" axis of its own, parallel with the earth's axis. The figure shows two of these private axes, one at 40° N latitude and one on the equator, as well the true axis protruding from the North Pole. The three axes are parallel; they are shown by the straight, solid north-pointing arrows.

Before continuing, note the meanings of *horizontal* and *vertical* in what follows, and see the inset in figure 4.4. A plane tangent to the earth's surface at any point is the horizontal surface at that point. Except at the poles, it is *not* parallel with the top and bottom edges of the page. A line through the point at right angles to the tangent plane is the vertical through the point. Except at the poles, it is *not* parallel with the left and right margins of the page. Keep this in mind as you read on.

Only at the pole itself, where the axis is vertical, is the rotating circular movement completely parallel with the ground, or horizontal. This movement is shown by the "twirls" (circles seen in perspective) above each axis.

At any site at an intermediate latitude (for instance, at 40° N as in the figure), the private axis emerges obliquely from the ground. It may be regarded as the resultant of two subaxes, shown in the figure as dashed arrows; one of the subaxes is the projection of the true axis onto the vertical at the site; the other subaxis is its projection onto the horizontal at the site. As the figure shows, the subaxes are shorter than the true axis, and the speed of rotation around each of them is less than around the true axis, as shown by the smaller twirls over the dashed arrows. These two circular movements at right angles to each other combine to give the true, oblique movement.

The private axis of an observer on the equator is horizontal. The circular motion around it is therefore confined to the vertical plane; its horizontal component is zero.

Now consider a pendulum suspended at each of the sites; imagine that each pendulum is supplied with just enough power to overcome friction and keep it swinging regularly; also, that it is hung so it can swivel freely around the point of suspension, ensuring that the plane of its swing will not rotate with the rotating earth but will remain fixed relative to the distant stars while the earth rotates beneath it. When the pendulum is at the North Pole, each swing will shift far enough to the right of the previous one to complete a full circle in exactly twenty-four hours. When the pendulum is at a middle latitude (for example, at 40° N, as in the figure), it will *tend* to be deflected by exactly the same amount in a plane at right angles to the earth's axis of rotation, that is, in a plane tilted obliquely to the horizontal. But gravity is strong enough to prevent any deflection in a vertical plane; the only deflection possible is the horizontal component, and this is less than the total that the earth's rotation "ought" to cause. The pendulum will therefore take considerably longer than twenty-four hours to complete a rotation relative to the ground. When it is on the equator, the pendulum will not rotate at all relative to the horizontal; its tendency—unrealizable because of gravity—will be to rotate wholly in a vertical plane. Such a pendulum, known as Foucault's pendulum,[10] was actually constructed in Paris in 1851, and its behavior gave incontrovertible evidence that the earth rotates on its axis; nowadays many museums have working replicas of it.

The argument shows how the earth's rotation sets up a "twist" affecting every point on earth except points exactly on the equator. The direction of the twist is the same everywhere and causes anything moving above the surface of the earth—the wind, a floating balloon, an airplane, a migrating goose, a swinging pendulum bob—to drift relative to the surface. To an observer on the earth, the drift appears rightward in the Northern Hemisphere and leftward in the Southern Hemisphere. Moreover, this drift, the Coriolis deflection, is greatest at the poles; at lower and lower latitudes, the horizontal component of the spin becomes less and less (the Coriolis deflection decreases); at the equator, the effect is zero and there is no deflection at all.[11]

## How the Winds Respond

Now back to the upper atmosphere winds, above the level at which friction with the ground causes complications (we come to those in chapter 5). Friction becomes negligible about 1 km above the earth's surface, and above that the troposphere continues for a long way; the height of the tropopause (the boundary layer separating troposphere and stratosphere) is about 10 km at the poles and more than 15 km at the equator. In describing events in the troposphere beyond the influence of friction, we are therefore considering a very thick layer—roughly, between 9 and 14 km thick—and conditions are not the same all through the layer.

First, recall figure 4.3, which shows the wind pattern as it would be *if* the atmospheric pressure decreased steadily from the equator to the poles and *if,* also, the earth did not rotate. These two "ifs"—simplifications—will now be abandoned.

Conditions leading to a wind pattern like that in figure 4.3 are unlikely to occur except near the top of the troposphere, and only occasionally even there. At lower levels the highest air pressures are seldom directly above the equator: more often there are two ridges of high pressure, one on each side of the equator. High in the troposphere, the ridges tend to be close to the equator and to each other. At progressively lower elevations they become farther and farther apart, being between 15° and 20° north and south of the equator at a height of 1 km; near ground level, they are usually at about 30° north and south and are known as the "subtropical highs."

This shows that figure 4.3 is an oversimplification of the wind pattern above a nonrotating earth except, sometimes, at the very top of the troposphere. At a lower level in the atmosphere, say at 5 km above the surface, the wind pattern if the earth did not rotate would be as shown in figure 4.5a. The highest pressures are at some distance from the equator, on each side of it. This causes the winds between the subtropical highs to blow toward the equator; poleward of the subtropical highs, the winds blow toward the respective poles. To repeat, this is a highly simplified version of what the wind pattern might be like *if* the earth did not rotate.

Now let the earth rotate, so that the Coriolis effect comes into play. The result is shown in figure 4.5b. Wind directions have turned through a right angle. In the Northern Hemisphere, what were south winds have become west winds or "westerlies" and what were north winds have become east winds or "easterlies," and vice versa in the Southern Hemisphere. (Recall that the name of a wind relates to the direction it is coming from: for instance, a west wind blows *from* west *to* east.) These winds are known as *geostrophic* winds; the term combines the Greek *geo-*, earth, and *strophe,* a turning. The reason the winds blow at right angles to the pressure gradient is as follows.

Consider a south wind in the Northern Hemisphere: it blows northward down the pressure gradient leading from the northern subtropical high to the low pressure area over the North Pole: this wind is deflected to the right (east) by the Coriolis effect. Were it not for the pressure gradient, the rightward deflection would turn the wind back on itself. The pressure gradient prevents this, however; the tendency of the wind to blow "downhill"—down the pressure gradient—and its tendency to turn right because of the Coriolis effect come into balance with the wind blowing directly across the pressure gradient, due eastward in this case. In maps showing the isobars (contours of equal atmospheric pressure) as well as wind directions, it is easy to see that the winds are parallel or nearly parallel to the isobars; examples are given in the following section.

The geostrophic winds are the dominant winds of the general atmospheric circulation, and it must be emphasized that they ultimately derive all their energy from the sun's heat, which produces atmospheric pressure gradients—

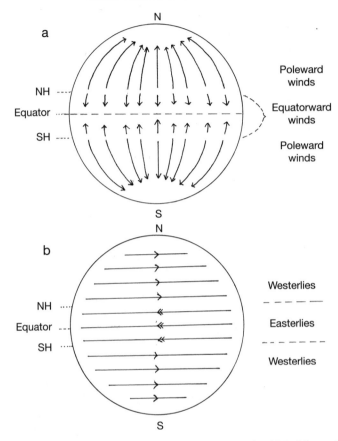

Figure 4.5. Idealized models of the winds at 5 km elevation, assuming (a) that the earth does not rotate and (b) that it does. NH and SH show the latitudes of the northern and southern subtropical highs.

the immediate cause of the wind. The Coriolis effect, caused by the rotation of the earth, determines the winds' directions, but it does not "drive" the winds in the sense of contributing energy to them.

## Jet Streams

The speed of the geostrophic wind at any point depends on the steepness of the pressure gradient and on the elevation. Consider elevation first. Low-level winds, at less than 1,000 m above the surface, are slowed by friction; above the friction layer, wind speeds continue to increase at increasing heights because of the decreasing density of the air. The fastest winds are at the level of the tropopause.[12]

The speeds of these winds also depend on the steepness of the pressure gradient, which is not the same at all latitudes. In the Northern Hemisphere the gradient is steepest at two "steps," one at about 60° N, on average, the other between 25° and 30° N, on average. (Note the words "on average"; the steps swing north and south over a wide range of latitudes, as we shall see.)

The fastest winds in the Northern Hemisphere therefore tend to be at the level of the tropopause in two widely separated latitude belts. These winds are the *jet streams*. The northern one is known as the polar jet; it blows at about 10 km above the ground; the southern one is the subtropical jet, which blows at a greater height—about 13 km up—because the tropopause is higher in the tropics than near the poles. Both are westerlies; the subtropical jet forms in the zone of westerlies (see fig. 4.5b), on the poleward side of the subtropical high.[13] A jet stream can be thought of as a current of air hurtling through the upper atmosphere at tremendous speed. In cross section it is shaped like a wide ribbon, hundreds of kilometers wide but only a few kilometers thick from top to bottom. The ribbon may be several thousand kilometers long.

The wind speed at the center of a jet stream is typically about 200 km/h, occasionally rising to over 450 km/h.[14] The kinetic energy of a 200 km/h wind blowing 10 km above the ground is no greater than that of a 113 km/h wind blowing at the surface, because the density of the air is less at high elevations than at low.[15] Even so, a jet stream is extremely powerful. When you see cirrus clouds ("mares' tails") drawn out into long, straight wisps far up in a blue sky, you are seeing clouds shaped by a jet stream. Even though the air may be calm at ground level, the evidence for a strong wind at great height is plain; the telltale clouds are seen most often in winter, when jet streams are strongest because the temperature contrast between the tropics and the polar regions is greatest.

*Rossby Waves*

The jet streams seldom blow "straight," in the sense of blowing steadily along a parallel of latitude; if they did, they would be little help in conveying the sun's heat from low latitudes to high. The polar jet usually follows a meandering course, blowing from the southwest and the northwest alternately. Figure 4.6 shows a typical path for it, snaking around the world in northern latitudes. The strongest winds follow this curving route in the same way that a

Figure 4.6 Representative map of the polar jet stream. The "waves" of its course are Rossby waves.

stream of water flows through a garden hose lying zigzag on the ground. The "waves" along its course are *Rossby waves*,[16] sometimes called *long waves*. Four or five of them, with wavelengths averaging 4,000 to 5,000 km, encircle the earth. The whole pattern of waves usually drifts slowly downwind (eastward) at a rate of about 4° of longitude a day, but this does not always happen; the pattern sometimes remains stationary for days on end and occasionally drifts backward (westward) for a while.[17]

Rossby waves form because high-pressure and low-pressure regions (highs and lows for short, or *anticyclones* and *cyclones*) develop in the upper atmosphere in the same way as they do at low levels, in response to differential heating at the surface. As a result, the high-level circulation at any one moment hardly ever has a pattern as simple as that in figure 4.5b, which is a simplified diagram corresponding, more or less, with the long-term average.

Consider how highs and lows affect the wind. What happens in the Northern Hemisphere is shown in figure 4.7. Figure 4.7a shows the isobars of an anticyclone, resembling the contours of a hill; if it were not for the rotation of

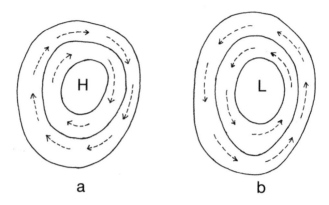

Figure 4.7 (a) An anticyclone and (b) a cyclone, both in the Northern Hemisphere, at an elevation above the friction layer at 1,000 m above the surface. The closed loops are isobars; the winds (arrows) are parallel to them.

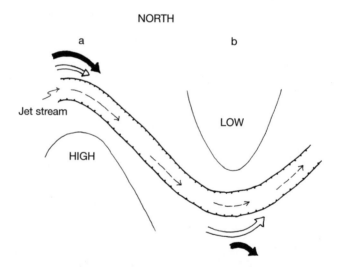

Figure 4.8. Details of one wave of the polar jet stream. Dashed arrows show the wind direction; the fine lines are isobars. (a) Here the jet blows around a ridge of high pressure, turning to the right. (b) Here it blows around a trough of low pressure, turning to the left. The solid and open arrows show the wave's two components: the solid arrows show the Coriolis effect as it would be in the absence of local pressure differences; it is stronger at high latitudes than at low. The open arrows show the wind directions resulting from the local pressure gradients, which happen to be steeper around the low than around the high.

the earth, causing the Coriolis effect, the winds would blow out radially in all directions "down the hill." Because of the Coriolis effect however, they are deflected to the right until Coriolis effect and pressure gradient are in equilibrium, at which stage they blow parallel with the isobars clockwise round the high. Figure 4.7b shows a cyclone; the pressure is lowest at the center, the isobars mimic the contours of a depression, the winds blow inward, "downhill," and deflection to the right turns them into counterclockwise winds.

Now consider how all this affects the flow of the polar jet stream. Visualize the jet in the subarctic as it approaches a ridge of high pressure extending northward from a localized anticyclone farther south (see fig. 4.8). The jet is deflected to the right (clockwise) by the anticyclone, boosting the Coriolis deflection already forcing it to blow toward the east. The enhanced deflection sends the jet southeastward, into lower latitudes. Sooner or later it encounters a trough of low pressure, which forces it leftward (counterclockwise) with sufficient strength to overcome its tendency to turn to the right. The jet now blows northeastward, toward an encounter with the next ridge of high pressure, where it will turn southeastward again. And so on, around the world.

In this way the jet stream carries cold air from polar latitudes toward the tropics and warm air back again, moderating temperature extremes worldwide.

# 5 ENERGY IN THE LOWER ATMOSPHERE

### THE WEATHER NEAR THE GROUND

*Contrasts between the Upper and Lower Atmosphere*

Wherever the wind blows, some of its energy is dissipated—converted to entropy—by the shearing of air against air; this happens at all elevations. The losses are far greater in the lowermost layer, however, because of friction with the surface—the drag of moving air as it passes across land or water. Drag also affects the direction of the wind, making atmospheric circulation far more complicated than it is aloft.

A number of other factors, too, have a stronger effect at low elevations than at high ones: the air temperature often varies greatly over small distances in response to the temperature at the surface, which may be a sunbaked desert, a cool forest, a cold lake, or the sea. Hills and mountains deflect the wind, funneling it through narrow valleys or forcing it up over high ridges. And there is more water vapor in the air at low elevations; most of the water vapor in the atmosphere is in the lowermost five kilometers.[1]

Daily fluctuations in air temperature are most pronounced in the layer heated directly by the surface. The air in the first few centimeters above the ground is heated by *conduction*, that is, by energy exchanges brought about by collisions between molecules of the air and of the ground. The warmed air becomes less dense and rises, conveying heat upward by *convection*. The layer warmed by conduction plus convection is seldom more than a few hundred meters thick. Air temperature at a greater height is not influenced by local surface temperature; it is controlled by more distant causes—the movement of air masses, and radiation from the sun and the ground.

The temperature of the air beyond the reach of surface effects is slow to react to changes in incoming radiation: it has high *thermal inertia*. The inertia is such that air temperature is scarcely affected by the daily variation of incoming sunlight caused by the earth's rotation.[2] Unless a new air mass invades an area, heralded by the arrival of a warm or cold *front* (the boundary between two air masses), daytime and nighttime air temperatures at a site do not differ significantly.

At first thought this is surprising: think of the chill of early morning in summer, just before sunrise while the grass is still drenched with dew. The chill affects only the air nearest the ground, however; you can perceive the marked temperature difference simply by noting that your face is warm while your feet feel frozen. The air temperature a mere hundred meters up may be the same as it was at sunset the previous evening.

In sum, conditions near the surface are much more variable, both spatially and temporally, than those higher up. Temperature, pressure, and humidity often change abruptly over short distances—in a word, energy becomes concentrated in confined areas—and changeable weather is the result.

## Surface Winds

The amount of heat stored in the atmosphere at any one time is about $1.3 \times 10^{24}$ J (joules).[3] Were it not for the constant movement of the air, this vast quantity of energy would be much less evenly distributed than it is. The energy itself makes the air move; that is, it causes the winds. As we saw in chapter 4, the total power of incoming solar radiation averages 340 W m$^{-2}$; of this

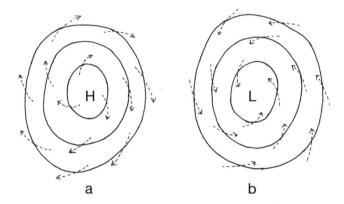

a                    b

Figure 5.1. Maps of (a) an anticyclone and (b) a cyclone, both in the Northern Hemisphere near the earth's surface. The closed loops are isobars. The winds (arrows) blow outward and clockwise from the center of the anticyclone; they blow inward and counterclockwise toward the center of a cyclone. Compare figure 4.7.

total, only about 2 W m⁻² is consumed in driving the winds, with more than half of it energizing the jet streams.[4] Low-level winds, however, are nearly as important as those aloft in transferring solar heat from the tropics to colder climes, and we now consider how friction with the surface (drag) affects the global wind pattern close to the surface.

Figures 4.5 and 4.7 showed how, at elevations above the friction layer, the winds blow parallel to the isobars. Their direction, at that level, is wholly determined by the pressure gradient and the Coriolis effect; in a word, upper-level winds are geostrophic. At lower elevations, where frictional drag is appreciable, the wind is governed by three factors: the pressure gradient, the Coriolis effect, and drag. Drag reduces the Coriolis effect, with the result that the wind is deflected through an angle of less than 90° from the direction of the pressure gradient.[5] The amount by which the Coriolis deflection is diminished depends on the nature of the surface over which the wind blows.[6] Over a calm sea, for example, the deflection is likely to be between 75° and 80°; over rough, hilly ground, between 50° and 55°.

The way this diminution of the Coriolis effect alters the winds blowing around anticyclones and cyclones (highs and lows, respectively) is shown in figure 5.1; compare it with figure 4.7. As figure 5.1 shows, the winds around an anticyclone blow *out*ward down the pressure gradient as well as clockwise;

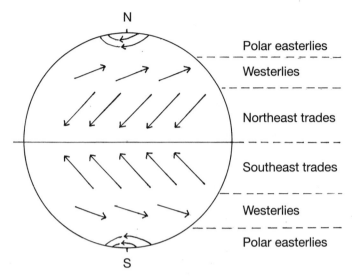

Figure 5.2. Idealized pattern of global wind circulation near the earth's surface, assuming there are no localized cyclones and anticyclones. Compare figure 4.5.

conversely, the winds around a cyclone blow *in*ward down the pressure gradient as well as counterclockwise; that is, they are not parallel with the isobars but are deflected to the right in both cases.

Now consider the effect on the global average wind pattern at the surface as it would be in the absence of the temporary, localized hot spots and cold spots that cause anticyclones and cyclones to form; this idealized pattern is shown in figure 5.2. Compare it with figure 4.5 and note how, in place of high-elevation westerlies and easterlies blowing along the parallels of latitude, the winds at the surface tend to blow across the latitudes. In this way air, and its thermal energy, is shifted from one latitude belt to another. The process is called *advection,* defined as the horizontal displacement of air and all its attributes. Note, too, that figure 5.2 shows zones of polar easterlies in both hemispheres that were not present in figure 4.5. The polar easterlies are caused by high pressures at a low elevation over the poles, where the air near the surface is cold and dense. The names "westerlies" and "easterlies" for the zones so labeled are inexact but traditional.

Figure 5.2 reveals a seeming paradox. The argument that cross-latitude winds transfer heat from warm regions to cold doesn't apply to the trade winds, which blow *toward* the equator in both hemispheres: they go the wrong way to equalize worldwide temperatures. In fact there is no paradox. Above every prevailing wind at the surface is a "return wind" higher up— there has to be, to prevent an impossible pileup wherever surface winds converge—and these high-level winds, especially the meandering jet streams, play a large part in transferring atmospheric heat poleward.

To say that low-level winds "return" at high elevations or, which comes to the same thing, that high-level winds "return" at low elevations (see fig. 4.3) amounts to saying that the air circulates vertically. Indeed it does, carrying its thermal energy with it.

### Vertical Movements of the Air

Look at figure 5.1 again. The winds blowing in toward the center of a cyclone converge; then, because the air must go somewhere, it rises. Likewise, the surface winds blowing outward from an anticyclone diverge, and air descends to fill the void. These upward and downward movements—up in a cyclone and down in an anticyclone—are too slow to be called winds. Vertically moving air usually moves at only one or two centimeters a second.[7]

To repeat: the air drifts *downward* into a surface anticyclone. This may seem like a paradox to anybody familiar with *thermals*, the warm air currents that rise from ground on a hot, sunny day when the barometer reads "high" and an anticyclone obviously prevails. The presence of thermals is revealed by the small, puffy cumulus clouds that often cap them, and also by soaring hawks and eagles using them for lift. Can their presence be squared with the general downward movement of the air in an anticyclone? It can: the thermals are scattered; their arrangement is like "the holes in the top of a pepper pot."[8] The rising thermals flow up as separate small airstreams through a mass of slowly descending air.

These various up and down movements show that the different layers of the atmosphere do not act independently; on the contrary, air and its thermal energy are continually exchanged between one layer and another, as well as shifting great distances horizontally.

### Water Vapor and Energy Transfers

A northerner feeling the warmth of a south wind is experiencing *sensible heat*. This surprising term simply means perceptible heat. The adjective is to differentiate it from *latent heat*, which is imperceptible. Two air masses can have the same temperature and feel the same, but if the concentration of water vapor in them is not the same, the moister air contains more latent heat per unit volume than does the drier air.

Latent heat is the heat liberated when water vapor condenses to liquid water or when liquid water turns to ice. Consider what happens when moist air cools and the water vapor in it condenses. At the outset, the water molecules are thoroughly mixed with all the other molecules of the air (mostly nitrogen and oxygen) and share in their constant motion. As the temperature drops, all the molecules slow down. Once the water molecules have slowed to a critical speed, they cling to any tiny particle they chance to bump into, which may be a dust mote, a smoke particle, a floating bacterium, a fragment of fly ash or the like, or any one of these that has picked up a few water molecules already; this is the way cloud droplets and fog droplets come into existence.[9] In uniting to form liquid water, the molecules slow down, abruptly losing energy. The energy cannot vanish: what the water loses is passed on to the air molecules, whose motion speeds up. That is, the temperature of the air rises: the heat that was latent in the water vapor  manifests itself as sensible heat— a rise in temperature.

Similarly, heat is released when cloud droplets freeze or when water vapor condenses directly into ice crystals without going through a liquid stage. Conversely, when water evaporates, ice or snow melts, or ice evaporates directly to vapor (sublimates), heat is absorbed; the air or ground that supplies the heat is cooled, and the heat itself becomes latent in the water vapor.

Condensation is extremely important in conveying heat from the surface to the atmosphere, which then carries it from one latitude to another. The hot, moist air of an equatorial rain forest, for example, carries both sensible heat and latent heat to cooler latitudes; the amount of latent heat it carries is about twice as great as the amount of sensible heat; it gains the latter by conduction and convection from the surface. In rainy, temperate latitudes, too, the air gains more latent heat than sensible heat; the opposite is true only in the subtropical deserts north and south of the equator. Over the oceans, the air acquires latent heat at a greater rate than sensible heat at all latitudes.[10] Notice that we are now considering the *rate* at which heat flows from land or sea into the air, called the *heat flux* and measured in watts per square meter.

Everybody experiences the release of latent heat from time to time, often without realizing it. Think of a clear summer evening when the temperature drops quickly after sunset because the ground radiates its heat out into space, promising a chilly dawn. If clouds gather, however, the temperature stops falling—it may even rise somewhat—and the night stays comfortably warm. This familiar scenario is usually attributed to the fact that the long wave radiation that had streamed from the earth into space when the sky was clear is absorbed by the clouds and radiated back to earth, keeping it warm. That is part of the explanation but not all of it: an appreciable fraction of the warmth is the latent heat of the water vapor in the atmosphere, released and made sensible as it condenses into clouds.


*Concentrated Energy: Storms*

Strong winds, heavy rain, lightning and thunder, and huge waves at sea all show that at times atmospheric energy becomes concentrated in confined areas. Why? The simple answer is that in a confined area something has happened to upset the normal equilibrium of calm weather. The next question— What sort of something?—has no single answer; it varies from one storm to another, depending on the kind of storm.

Most storms can be classed as cyclones of one sort or another: they are accompanied by winds blowing counterclockwise in the Northern Hemisphere and clockwise in the Southern Hemisphere. They are one of two strongly contrasted kinds of cyclones, however, depending on the latitude: *tropical cyclones* differ radically from *midlatitude cyclones.* The most intense tropical cyclones occur over the ocean: those in the western Atlantic are called *hurricanes,* and those in the western Pacific are *typhoons.*

Tropical cyclones and midlatitude cyclones derive their energy from entirely different sources.[11] Water vapor is abundant in the air over warm ocean currents, and the latent heat of the vapor, released when it condenses, is the energy source for tropical cyclones. The strongest of them also differ from midlatitude cyclones in having an "eye" at the center, where a current of warm air flows downward; the eye shows as a dark spot at the center of the swirl of white clouds in many satellite photos of hurricanes. But the central current of air in midlatitude cyclones flows upward.

The energy source for midlatitude cyclones develops when two air masses at different temperatures come to be side by side; a horizontal temperature contrast results, which is intrinsically unstable. The separation of warm, light air from heavy, cold air by a vertical surface (a front) creates potential energy (PE). The greater the temperature contrast, the greater the energy. Some of this energy is released, that is, converted to kinetic energy (KE), by a re-

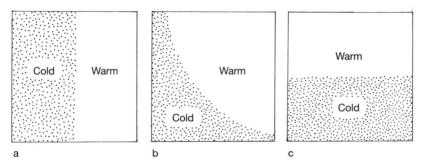

Figure 5.3. Side view of two adjacent air masses. (a) A cold air mass has drifted into contact with a warm air mass, and they are separated by a vertical front; the arrangement is unstable (possesses high available PE) because the warm air is less dense than the cold. (b) The cold air has started flowing under the warm air. (c) The final arrangement, with the cold air wholly below the warm air; the arrangement still possesses PE relative to the ground surface, but none of it is available.

arrangement of the air masses. The cold air, because of its greater density, flows under the warm air (see fig. 5.3); the flow, down a pressure gradient, undergoes Coriolis deflection, and the result is a cyclone.

After the repositioning of the air masses, the surface separating them is horizontal. The upper layer still has PE relative to the ground, in the same way that a boulder sitting in the middle of a level plateau has PE relative to sea level (see chapter 2), but it is unavailable—there is no way for it to be converted to KE. Because of this, the efficiency of the conversion of PE to KE is low, only about 1 percent.[12]

The energy released, from start to finish, by a typical midlatitude cyclone averages about $2 \times 10^{13}$ MJ (megajoules), and that of a tropical cyclone about $2 \times 10^{11}$ MJ, only one-hundredth as much. The average area of a midlatitude cyclone is about two hundred times that of a tropical cyclone, however, so that the energy *per unit area* is actually about twice as great in the tropical cyclone. Tropical cyclones are the most energetic of all storms. If we assign a rating of 100 percent to the energy per unit area of an average tropical cyclone, the average energy per unit area of lesser storms is approximately as follows: a tornado, 80 percent; a midlatitude cyclone, 50 percent; a dust devil, 8 percent; and a severe thunderstorm, 6 percent.[13]

Solar energy is transported from the tropics to the polar regions by ocean currents and winds acting in combination. The rate of transport is greatest at about latitudes 30° N and 30° S, where the two modes of transport are approximately equal. Between these latitudes, in the tropics and warmer subtropics, more energy is transported by ocean currents than by the atmosphere, whereas poleward of them atmospheric transport dominates.[14]

Violent winds—sometimes more than 500 km/h—are not the only displays of nature's energy that storms provide. Lightning and thunder from electrical storms are equally awe inspiring. We consider them in chapter 16.

*How Atmospheric Energy Is Dissipated*

The atmosphere is gaining solar energy all the time, at a rate of almost 220 W m$^{-2}$. The earth as a whole receives solar energy at a rate of 340 W m$^{-2}$. As figure 4.1 showed, 30 percent of this total is reflected back into space by the atmosphere and 6 percent by the ground (land and sea); the remaining 64 percent, nearly 220 W m$^{-2}$, provides the energy that is held, temporarily, in the atmosphere. It does not accumulate; if it did, the temperature would go on rising indefinitely. Therefore it must somehow be dissipated.

The atmosphere holds energy in four forms.[15] First in importance is its internal (thermal) energy, the energy of its molecules' never-ending random motion. Second is the potential energy resulting from the temperature differences within the atmosphere (see the preceding section). Third is the latent energy of the water vapor in the atmosphere, ready to liberate heat when it condenses. Fourth, and least in quantity at any one moment, is the kinetic energy of moving masses of air—winds.

All this energy is constantly being augmented by incoming solar radiation and must be continuously dissipated at the same rate if equilibrium is to be maintained. The absorption of solar radiation increases the total internal energy of the atmosphere and also, because of unequal heating, its potential energy. Most of this energy is dissipated as fast as it is absorbed, by reradiation into space in the form of long wave (infrared) radiation.

About 30 percent of the absorbed solar energy evaporates water, both from the sea and from freshwater lakes and rivers. The latent energy in the vapor turns back into heat when the vapor condenses to raindrops; the raindrops warm the air and the ground, and the energy is ultimately lost to space as long wave radiation.

The wind does much of its work on the sea. A large proportion of its energy is used in driving ocean currents and raising waves. Some of this energy is transferred to the water as kinetic energy while some, as always, is lost as "waste" heat—that is, entropy (see chapter 3).

The wind also does the work of wind erosion—the natural sandblasting that shapes rocks and cliffs in regions with dry climates; erosion entails friction, hence more waste heat. Indeed, whenever the wind shifts anything—when it builds a sand dune, knocks down a tree, or blows your hat off—the entropy of the universe is increased by a tiny amount. Much wind energy is dissipated as the heat produced by viscous drag because of shearing within the wind itself.

A very small fraction of the wind's work consists in turning humanity's windmills. The output is sometimes milled flour, plus of course the inevitable waste heat; or it may be electrical energy from a generator driven by windmills. For more on the latter topic, see chapter 19.

*The Energy in a Rainstorm*

Back to the topic of rain. Rain provides an opportunity to demonstrate with actual numbers what becomes of the energy in a common meteorological event.

A heavy rainstorm obviously holds considerable energy. So, what becomes of the energy? It is easy to calculate the work done when a measured depth of rain falls on a known area of land from a cloud at known height. (At least it's easy provided you can assume that the air was saturated with moisture so that none of the rain evaporated as it fell, as it always does in relatively dry air.) Suppose, for example, that 10 mm of rain falls on one hectare (10,000 m²) of land from a cloud 200 m up. Take the density of the rainwater to be 1,000 kg m⁻³. The volume of the rainfall is

$$\text{depth} \times \text{area} = 0.01 \text{ m} \times 10,000 \text{ m}^2 = 100 \text{ m}^3,$$

and its mass is

$$\text{volume} \times \text{density} = 100 \text{ m}^3 \times 1,000 \text{ kg m}^{-3} = 100,000 \text{ kg}.$$

It follows (see chapter 2) that the work done, or energy used up, as the rain falls to the ground from a height of 200 m is

$$\text{mass} \times \text{acceleration due to gravity } (9.81 \text{ m s}^{-2}) \times \text{distance fallen}$$
$$= 100,000 \text{ kg} \times 9.81 \text{ m s}^{-2} \times 200 \text{ m} = 1.962 \times 10^8 \text{ J}.$$

The computation is straightforward, but where have the $1.962 \times 10^8$ J gone once the rain has reached the ground? They cannot have vanished, so what has become of them? They have been dissipated, partly in the air and partly at the point of impact with the surface (land or sea).

Consider the details of what happens: as the raindrops fall, they are slowed by the drag between each drop and the air. Indeed, a falling raindrop cannot fall faster than its *terminal velocity*, the velocity at which the pull of gravity is exactly canceled out by the drag. For a raindrop 2 mm in diameter (a typical size), the terminal velocity is 6.5 meters per second;[16] it reaches this speed after falling a mere 2 m. Once a falling drop reaches its terminal velocity it cannot speed up any more. If it were not for drag—if the drops were falling in a vacuum—their velocity after a fall of 200 m would be 62.64 m s$^{-1}$, almost ten times as great.[17] The kinetic energy (KE) of the falling rain when it hits the ground is (see chapter three)

$$\tfrac{1}{2} \times mass \times velocity^2 = \tfrac{1}{2} \times 100{,}000 \text{ kg} \times (6.5 \text{ m s}^{-1})^2 \text{ J} = 2.1125 \times 10^6 \text{ J}.$$

This is a mere one-hundredth of the KE it would have possessed had there been no drag. In the no-drag case, the KE would have been

$$\tfrac{1}{2} \times 100{,}000 \text{ kg} \times (62.64 \text{ m s}^{-1})^2 \text{ J} = 1.962 \times 10^8 \text{ J},$$

as we found above in computing the work done by the falling rain.

The rain has indeed done this much work; all but $1.94 \times 10^8$ J, however, has been dissipated as heat on the way down. In practice, the rise in temperature (it would be 2°C at most) is masked because the air is simultaneously losing heat to the cold raindrops, falling from cold air above. The $2.1125 \times 10^6$ J that the rain still has left on reaching the ground remains to be disposed of. It no more vanishes than the rain itself vanishes when it soaks into the ground or becomes one with the water in the sea or a lake: it only seems to vanish. Consider a single raindrop. If it strikes water, or a rock, it creates a small splash, and the drop itself is deformed. If it strikes soil (or soft clay, or sand) it makes a tiny dent, and both the raindrop and the surface struck are deformed. These events—splash, deformation—consume each raindrop's remaining energy, and the account is precisely balanced at last.

# 6  THE SUN, THE WIND, AND THE SEA

## The Sun and the Sea

When sunlight shines on the sea, what becomes of it? Some is reflected back into the sky, and the rest penetrates the surface to contribute its energy to the water; in a word, it is absorbed. This statement prompts two questions. What are the proportions reflected and absorbed? And what becomes of the absorbed energy?

Consider the first question. Recall the account in chapter 4 of the solar energy budget for the earth as a whole. It is believed (see fig. 4.1) that, averaging over all seasons, approximately 70 percent of all incoming sunlight is reflected back into space. This quantity—70 percent—is the earth's reflectivity or, to use the technical term, its *albedo*. Seventy percent is an average for the whole earth, but different surfaces have markedly different albedos. For example, the albedo of new-fallen snow is extremely high, up to 95 percent; that of calm water (this includes the ocean), is occasionally as low as 2 percent, rising to about 10

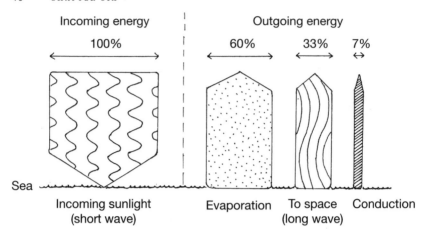

Figure 6.1. The fate of the short wave solar radiation absorbed by the sea (downward-point-ing arrow on the left). The arrows pointing upward on the right show the proportions of in-coming energy dissipated in various ways. See text for details.

percent for a rough sea; that of clouds is anywhere between 30 and 90 percent, depending on the type and thickness of the clouds.[1]

The low albedo of the ocean, ranging from 2 to 10 percent, doesn't mean, however, that the oceans always absorb from 90 to 98 percent of incoming solar radiation: far from it. The sunlight is often intercepted by strongly reflective clouds in the atmosphere before it ever reaches the ocean. Indeed, cloudiness probably has a stronger influence on an area's albedo than any other factor.[2] Thus when you look over a calm, blue, sunlit sea, it is absorbing nearly all of the solar energy falling on it; this is especially true at low latitudes, where the sun's rays are nearly vertical at midday; but when you look at a gray, storm-tossed sea beneath a gray sky covered with thick clouds, the sea may be ab-sorbing only 10 percent or less of the energy reaching the cloud tops.

Now for the fate of the absorbed solar energy, which is augmented to a neg-ligible degree by heat from the earth's interior, and also by the frictional heat generated by the movements of seawater, especially surf. Figure 6.1 shows what happens to it.[3] All the energy must ultimately be returned to space; if it were not, the oceans would heat up. Therefore the incoming energy balances the outgoing. More than half of it is disposed of by evaporative cooling; sec-ond in importance is heat loss in the form of long wave (infrared) radiation to space; last, a small amount is lost by conduction to the air in contact with the sea surface. Typical proportions are shown in the figure.

Some solar energy is absorbed only a short distance below the ocean surface, by living things—green plankton carrying on photosynthesis. This energy is subsequently converted to heat by living organisms as they breathe and by dead ones as they decompose, warming the water by an imperceptible amount. The amount of energy used and dissipated by marine life is too small to be singled out for special notice in figure 6.1. It forms a tiny fraction of the three "outgoing" arrows on the right.

Evaporative cooling entails the withdrawal of thermal energy from the seawater and its conversion into latent heat in the water vapor that forms over the sea (see chapter 5). Occasionally energy travels the other way: water vapor in the air condenses to water at the surface of the sea, liberating latent heat that warms the water. But on average the net transfer of latent energy is from the sea to the air.

Likewise, heat can be conducted in either direction, depending on whether the sea at the surface is warmer or cooler than the air in contact with it. In other words, warmth—thermal energy—can be transferred downward from warm air to cool sea, or upward from warm sea to cool air, by conduction. In the latter case convection speeds up the transference, provided the temperature of the air falls at progressively greater heights above the surface; if it does not—if there is an inversion (the air temperature rising at greater heights)—convection can carry the warmed air only a short distance up. In any case the net transference of heat is upward, making conduction one of the ways—the least important way—the sea dissipates heat.[4]

The warmth of the sea is also transported horizontally, by currents; as with similar shifts of thermal energy in the atmosphere, such horizontal movements are called *advection*. Advection is not shown in figure 4.1 because directions and distances vary tremendously from place to place. All that can be said is that, over time, the advective heat transfers at any one location always cancel each other out.

*How Sunlight Penetrates the Sea*

As a whole, the ocean absorbs far more solar energy than does the land. This is scarcely surprising when you consider that the earth's surface is 71 percent ocean and 29 percent land. The ocean also absorbs more than the land per unit area, because of its lower albedo. Seawater is much less transparent than the atmosphere, however, and this limits the depth to which sunlight can penetrate.

Seawater also filters the sunlight shining through it.[5] The water is more transparent to blue light than to other colors of the visible spectrum and almost opaque to invisible radiation, both infrared and ultraviolet. More than half the incident sunlight, including all of its infrared component, is absorbed in the topmost meter of the sea. By 10 m down, all colors but blue are gone. This is the reason the sea looks blue: the color comes from the light back-scattered by the molecules of seawater, and because all other colors are absorbed in the topmost layers, blue is the only one available for back-scattering from lower down. Reflection of the blue sky above makes a comparatively minor contribution to the blue of the sea on a sunny day.

Biologically productive water, containing mineral particles, biological pigments, and plankton, is much less transparent than pure water and usually looks greener.

The solar radiation shining into the sea warms it, making the surface water warmer than the water below it. This arrangement is very stable, in strong contrast with the state of affairs in the atmosphere, where sunshine heats the surface, which in turn heats the air in contact with it. This places warm air below cool air and leads to *in*stability: convection currents of air—thermals—rise up from the hot ground, and in extreme cases, large air masses are overturned.

The way sunlit seawater heats up is strongly affected by its transparency.[6] On a hot summer day the sun could, in theory, raise the temperature of the topmost 5 cm of clean seawater by 40°C. It doesn't happen because, during daylight, the sea is losing heat almost as fast as it gains it. It loses it by radiation, most rapidly from the very topmost skin of water, about 0.05 mm thick. The temperature within this skin is less at the top than at the bottom (in contrast to the state of affairs just below it), causing the thin skin of water to mix convectively and cool down; at the same time, turbulence in the air over the water induces matching turbulence, hence mixing, in the water itself. The rapid cooling prevents the theoretically possible (and incredible!) gain of 40°C that could happen were it not for the cooling; in practice, the temperature of the sea surface seldom changes by more than 1°C between day and night. The sea doesn't cool appreciably after sunset, whereas the atmosphere is cooling rapidly. It is this that makes the sea seem so warm when you swim at night; you are enjoying the stored energy of sunshine.

*Movement in the Sea*

The sea is never still. It moves in a variety of ways, from a number of different causes. Waves and tides are the movements most familiar to land dwellers, and they are the topic of chapters 7 and 8. In this chapter we consider ocean currents, the "winds" of the sea.

In the open ocean, far from any coastlines, the strongest ocean currents are those at the surface, caused by winds. Weaker currents at greater depths have a variety of causes. Last, some currents at the bottom of the ocean, known as *hydrothermal currents,* are caused by heat from the earth's interior; they are discussed in chapter 15.

Waves as well as surface currents are driven by wind energy. When the wind gets up over a calm sea, it drags along a thin skin of surface water, simultaneously starting a shallow current and creating tiny ripples. The ripples catch the wind and become small waves; this increases the wind drag, so that the waves continue growing in a feedback process. The higher the waves, the deeper the troughs between them, so the more deeply the wind can reach into the water and the thicker the current layer becomes. Despite appearances, the winds spend considerably more energy driving currents than they do raising waves.[7]

Although ocean currents resemble atmospheric winds in many ways, there are striking contrasts between them. Take the matter of speed: one writer has said that "in nature, air flows are normally about fifteen times as rapid as flows of water. Thirty meters per second is a hurricane of air; two meters per second is a torrent of water."[8] At the slow end of the scale, the speeds at which flow is perceptible without instruments are in about the same ratio: the threshold of perceptibility is about 30 cm s$^{-1}$ for an air current and about 2 cm s$^{-1}$ for a water current.

The difference is due to the different viscosities of air and water.[9] The viscosity of a material—be it a gas, a liquid, or a paste—is a measure of the force required to maintain a shearing motion of given speed in the material. To take a culinary example, envision an ice-cream sandwich, consisting of a flat slab of ice cream between two wafers. Suppose the bottom wafer is held firm on a horizontal surface; now you push the top wafer horizontally until it is sliding at a constant speed. For an ice-cream slab of given dimensions, the speed of the shearing motion depends only on the force you apply and the ice cream's viscosity: knowing the speed, the force, and the slab's dimensions, you can compute the viscosity of the shearing movement in the ice cream. As you would expect, it is high when the ice cream is frozen hard and low when it is soft.

Let's return to the contrast between air and water. The viscosity of seawater is about sixty times as great as that of air at sea level.[10] No wonder their typical speeds of flow are so different. Their kinetic energies per unit volume are different too. For example, here are the speeds and energies per unit volume of three representative ocean currents:[11]

| Current | Speed (m s$^{-1}$) | KE (J m$^{-3}$) |
| --- | --- | --- |
| Florida Current (extremely fast) | 2.2 | 2,480 |
| Gulf Stream off Europe (slow) | 0.1 | 5 |
| Just strong enough to feel | 0.02 | 0.2 |

And here, for comparison, are the speeds and energies of representative winds:

| Wind[12] | Speed (m s$^{-1}$) | KE (J m$^{-3}$) |
| --- | --- | --- |
| Violent storm (great damage) | 30.0 | 540 |
| Light breeze (leaves rustle) | 2.0 | 2.4 |
| Just strong enough to feel | 0.3 | 0.05 |

As you can see, ocean currents have a higher kinetic energy per unit volume than comparable winds; this is because of the much greater density of water. The densities of seawater, and of air at sea level, in kilograms per cubic meter, are 1,024 and 1.2, respectively; that is, seawater is about 850 times as dense as air.[13] Note, too, that the energy in ocean currents is far more concentrated. As we saw in chapters 4 and 5, wind speeds increase with height above the ground, so that air at all levels is in motion. In contrast, the speed of ocean currents decreases rapidly as you move downward from the surface; as we shall see later, nearly all the wind-driven movement is confined to a shallow surface layer, ranging from 100 m to 500 m thick.

### Wind-Driven Currents and the Ekman Spirals

We have noted already that the strongest ocean currents are those at the surface, powered by the wind. The proof that the wind controls them is that they respond within a few hours to changes in wind direction.[14]

The speed of a wind-driven current is about 3 percent or less of the speed of the wind driving it. Much of the wind's energy is used up in the friction (strictly speaking, viscous drag) by which the wind drags the water along. Drag between wind and sea at the ocean surface and between layers of water

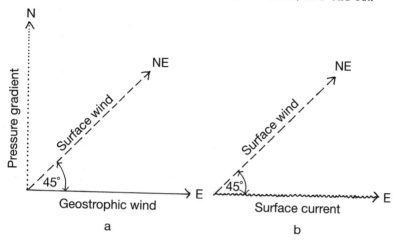

Figure 6.2. Map of how the winds above the ocean, and the surface current, respond to the Coriolis effect in the Northern Hemisphere: (a) atmospheric winds; (b) the wind and the current at the surface. See text for details.

below the surface converts much of the mechanical energy of winds and currents to waste heat—entropy—an item not to be omitted from a complete energy budget.

The direction of a wind-driven current does not coincide with that of the wind driving it. This is another manifestation of the Coriolis effect described in chapter 5, which causes a surface current to be deflected through an angle of about 45° to the right of the wind. An angle of exactly 45°, which is what the simplest mathematical model of ocean circulation predicts, is not to be expected in the real world; the simple model assumes that the only factors determining a current's direction are the wind and the earth's rotation, whereas in reality a number of other factors influence the outcome. Assuming the simple model to be correct, figure 6.2 compares what happens in the lower atmosphere with what happens at the sea surface (we consider later what happens just *below* the surface). The figure applies to the Northern Hemisphere; as always happens with processes dependent on the earth's rotation, the pattern of events in the Southern Hemisphere is the mirror image of those in the Northern Hemisphere.

Figure 6.2a shows events in the atmosphere. The dotted arrow pointing north shows the direction the wind would blow given a nonrotating earth and a gradient of pressure decreasing to the north as it usually does. The solid

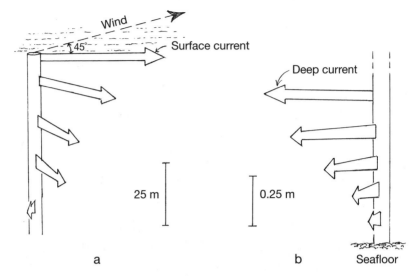

Figure 6.3. Ekman spirals at (a) the surface and (b) the bottom of the ocean. In both diagrams the open arrows low down point upward from the page. Note the different scales.

arrow pointing east is the geostrophic wind (see chapter 4), deflected 90° to the right by the Coriolis effect acting in the absence of drag. The dashed arrow pointing northeast shows the wind at the sea surface, where the Coriolis effect *has* been diminished by drag (see chapter 5); only half the deflection that produced the geostrophic wind remains, which makes it appear, as you descend from the heights, that the geostrophic wind has been deflected 45° to the left.

This is the wind that drives the current directly below it, as shown in figure 6.2b, where we see that the current at the surface (wavy arrow) has been deflected 45° to the right of the wind driving it. Behold, the surface current flows in the same direction as the geostrophic wind thousands of meters overhead, above the atmosphere's friction layer.

The next question is, How does the current flow at progressively greater depths below the water surface? The answer is given in figure 6.3a, which shows the directions and speeds of currents forming the *Ekman spiral,* so called in honor of the oceanographer who developed the first mathematical model of ocean circulation.[15] The figure shows the water as a stack of horizontal layers, with the current in each layer moving more slowly than that in the layer above and somewhat to the right of it because of Coriolis deflection; in practice, of course, the layers are infinitesimally thin and blend into each other. The current peters out gradually, spiraling clockwise all the while. At the depth where its direction is directly opposite to the direction at the surface, known as the *Ekman depth*, the speed is only 4 percent of what it was at the surface.[16] The water between the surface and the Ekman depth is the *Ekman layer*, which corresponds to the friction layer in the atmosphere.

The thickness of the Ekman layer depends on the wind speed and the latitude: the higher the latitude, the thinner the Ekman layer. For example, at latitudes 10°, 45°, and 80° (in either hemisphere), the Ekman depths would be 100 m, 50 m, and 45 m, respectively, given a fresh breeze of 10 m s$^{-1}$, that is, a wind strong enough to raise moderate waves with many whitecaps.[17] Doubling the wind speed to 20 m s$^{-1}$ (a fresh gale) doubles the Ekman depths. Clearly, the Ekman layer is extremely thin compared with the friction layer of the atmosphere.

Now notice what seems at first sight a surprising fact: as you descend from high in the atmosphere, wind direction twists to the left;[18] but as you descend from the sea surface into the depths, current direction twists to the right. Why the difference? Here is the answer. When the Coriolis effect twists the direction of a current of air or water, it twists it to the right of the direction in which it is being driven by the action of an external force, as explained in chapter 4. In the atmosphere, the external force is the atmospheric pressure gradient, exerting a force to the north in figure 6.2a (dotted arrow). In figure 6.2b the driving force is the wind, exerting a force to the northeast. Looked at in this light, it is clear that there is no difference; the two halves of figure 6.2 can be seen to correspond if you conceal the geostrophic wind arrow in 6.2a; then, in the two diagrams, you see arrows showing the initial causes and final outcomes of, respectively, a pressure gradient in the atmosphere and a wind driving the sea surface. In both cases the effect is directed 45° to the right of the cause.

Because of the constantly turning current direction as you descend below the sea surface, it follows that the bulk of the water is not moving in the same direction as the surface current; in fact the average direction of flow is at right angles to the wind direction, equivalently 45° to the right of the current at the surface. The average current is called *Ekman transport*; it is the flow that matters when we come to consider the transport of thermal energy by the ocean, though it is of no importance to sailors.

Other Ekman spirals form at the bottom of the ocean, where slow deepwater currents are finally braked to a stop by the drag of the seafloor (fig. 6.3b). Because the currents are weak, the Ekman layers are thin; at 45° latitude and a current speed of 0.1 m s⁻¹, the top of the Ekman layer is only 50 cm above the seafloor. The direction of a sea-bottom spiral matches the direction of the atmospheric Ekman spiral; that is, it twists to the left. This is as you would expect. In both cases, a geostrophic current of air (fig. 6.2a) or water (fig. 6.3b) is slowed by friction with a surface below it.

Returning to the Ekman layer at the top of the ocean, it is the layer in which virtually all wind-driven currents flow ("virtually" because, according to the model, wind-driven currents still have 4 percent of their strength at the Ekman depth).

The bulk of the energy imparted to the sea by the winds is used up in the surface Ekman layer. It is dissipated by viscous shearing as layers of water slide over each other. Viscosity exists in two forms: *molecular viscosity* and *eddy viscosity*.[19] In molecular viscosity, individual molecules of water pass up and down between adjacent layers of water: then, whenever a fast molecule moves down among the slower molecules in the layer below it, it is slowed by collisions with the slow molecules and at the same time imparts some of its speed to them. Conversely, whenever a slow molecule moves up among the faster molecules in a layer above it, it is speeded up by collisions with the fast molecules, slowing them in the process. In this way the speed differences are evened out. In eddy viscosity the mechanism is the same, but the objects exchanging energy are big chunks of water instead of individual molecules. Eddy viscosity dissipates from $10^7$ to $10^{11}$ times as much energy as molecular viscosity does. Eddy viscosity is therefore vastly more important in slowing ocean currents.[20] Realizing this, in 1902 Ekman succeeded in developing the first useful model of ocean circulation. Practically all the more recent models are refinements of Ekman's model made possible by high-speed computers.

## Hills and Dales of the Ocean's Surface

Disregarding the small-scale ups and down of the waves, the surface of the ocean is not horizontal; in other words, sea level is *not* level, in spite of appearances.[21] A three-dimensional model of an ocean, with an enormously exaggerated vertical scale, would show a surface of smoothly sloping hills and valleys. The slopes are much too gentle to be directly observed from the surface, and their existence can only be inferred, usually from satellite observations. Their presence shows that gravitational potential energy (PE) is to be found at sea as well as on land, though in comparatively tiny amounts. Just as a rock at the top of a precipice has PE relative to the lowlands below, the water

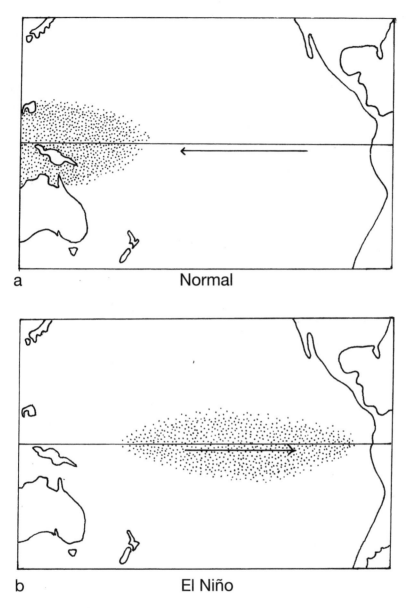

a           Normal

b           El Niño

Figure 6.4. The Pacific Ocean (a) under normal conditions and (b) during an El Niño event. Stippling shows warm water (about 30°C or more); arrows show currents.

atop a "sea hill" has PE relative to the lower surfaces surrounding it; the PE is converted to KE when water flows down the hill.

The tides are the most obvious cause of "hills" and "valleys" at sea. This follows directly from common sense, with no need for satellite data. The tides rise and fall because the whole body of ocean water encasing the earth, except where continents protrude, is not a spherical shell but a more or less football-shaped shell. The tides rise and fall because two bulges of water on opposite sides of the earth—the ends of the football—rotate around the earth. The bulges are low hills of water, each having a basal diameter the same as the earth's diameter. If you watch the tide rising at the beach, you are seeing one of these hills coming toward you; and if you begin your vigil at water's edge at low tide and don't get out of the way, the hill engulfs you. The evidence that the sea surface slopes could hardly be more convincing. (For more on tides, see chapter 8.)

Smaller hills form, superimposed on the tidal hills. A common cause is a fall in atmospheric pressure; if the pressure drops by 3 percent—equivalent to a drop from "set fair" to "change" on a household barometer—the sea level rises by about 30 cm.[22] A rise or fall in sea level accompanying a fall or rise (respectively) in air pressure is known as the *inverted barometer effect*. The flooding that often accompanies severe hurricanes is a manifestation of the effect.

The sea also piles up where a current is halted by a barrier. For example, sea level in the tropical western Pacific is normally higher than in the tropical eastern Pacific because the westward-flowing South Equatorial Current piles up on reaching Indonesia and northern Australia. On the eastern side of the Pacific, cool water flows into equatorial latitudes from the south to replace that flowing west, causing a marked temperature gradient as well as a slope; the result is heaped-up warm water in the west and cool water at a lower level in the east; the difference in elevation across the width of the Pacific is about 50 cm. But every few years, air pressure rises over the western Pacific and falls over the eastern Pacific; the trade winds weaken, and the piled-up warm water in the west flows back into the eastern Pacific, raising surface temperatures by 8 or 9°C (fig. 6.4). This is an *El Niño* event; it is a rearrangement of thermal energy on a geographic scale, and it causes climatic havoc.

Currents also cause water to pile up even in the absence of land barriers. Indeed, wherever currents flow, the ocean surface slopes. The slopes are imperceptible: for instance, at 45° latitude, a current of 1 m s$^{-1}$ causes a vertical difference in sea level of about 1 m in every 100 km, that is, a slope of 1 in 100,000.

The broadest "hills" in the ocean are enclosed in the huge "ring" currents known as *gyres*. As examples, figure 6.5 shows the gyres in the northern and

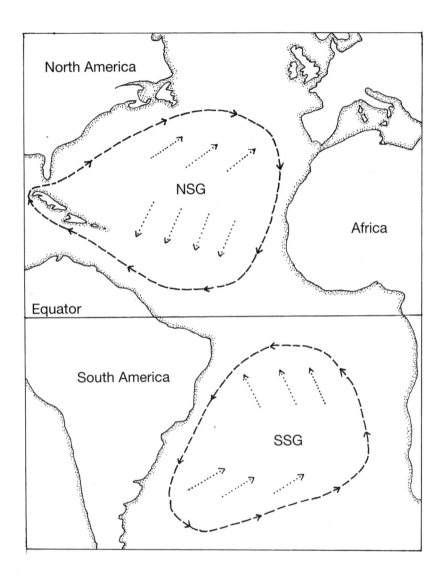

Figure 6.5. The Atlantic Ocean, showing its northern and southern subtropical gyres, NSG and SSG, and the currents circulating around them. (Other currents are not shown.) The dotted arrows show the prevailing winds: northeast and southeast trades on either side of the equator and "westerlies" at higher latitudes in both hemispheres.

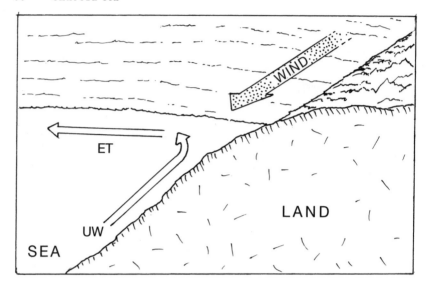

Figure 6.6. An upwelling. The view is northward, into the prevailing wind (stippled arrow), along the west coast of California. The hollow arrows in the sea show Ekman transport (ET), powered by the wind, and an upwelling (UW) dragged up from deeper water. The slope of the sea surface is exaggerated.

southern Atlantic. The currents are driven by the winds and deflected by the landmasses. But that is only part of the explanation for them and doesn't account for their constancy: additional mechanisms, with feedback, are at work. Coriolis deflection continuously turns the currents, and it acts more strongly on warm water at the surface than on the cooler water below because the warm water is less dense. As a result, warm water piles up to form a hill within the gyre;[23] in the North Atlantic, the surface at the summit of the hill is about 1.5 m higher than the surface at the periphery.

The water on the "hillslopes" has potential energy by virtue of altitude, small though it is. Water flows downslope under the pull of gravity and is deflected because of the Coriolis effect until it is flowing very gently downhill almost parallel with the contours—clockwise in the Northern Hemisphere, counterclockwise in the Southern Hemisphere. It behaves as the air does around an atmospheric anticyclone. The currents around the "hill" are almost geostrophic (controlled solely by gravity and the Coriolis effect) because the drag is very slight. They reinforce the wind-driven currents that were their initial cause.

Because the currents around gyres are maintained by two processes acting in concert, they are very constant. They do not respond nearly as quickly to changes in wind direction as do currents driven wholly by wind. Energy to keep the currents flowing is stored, as potential energy, in the topography of the water surface. It is released—converted to kinetic energy—gradually over many days. This makes the currents immune to short-lived wind changes.

Valleys as well as hills contribute to the relief of the sea surface. They appear where surface currents diverge from each other because the winds driving them diverge. Subsurface water flows upward to make good the loss.

Slopes also form where a current flows away from a coastline. This happens, in either hemisphere, wherever the prevailing wind blows toward the equator beside the west coast of a continent (see fig. 6.6). The wind drives a current and, as always, the Coriolis effect goes to work. The average flow is Ekman transport moving the bulk of the water at right angles to the wind direction—directly away from the land. The water surface develops a slope going uphill away from the land, and an ascending current of water from the depths flows upward to take the place of the water that Ekman transport has removed. This is an *upwelling*. The upwelling water is cooler than surface water, but not really cold: it rises gently, at about 10 m per day, from a shallow depth of 300 m at most.

Upwelling explains the surprising coolness of inshore waters off the west coasts of North and South America and southwestern Africa. Californians in particular are well acquainted with upwelling: along the shore at Cape Mendocino, the surface temperature of the water in August is about 7°C lower than the temperature 1,500 km out to sea at the same latitude.[24]

*Density Currents in the Depths*

Up to this point we have been concerned chiefly with the surface layer of the oceans, where wind-caused currents predominate. The layer is from 100 m to 500 m thick and is called the *mixed layer* because it is continually stirred by currents, including those that flow upward and downward. The mixed layer contains only about 2 percent of the whole ocean.

Below it, beyond the reach of the winds, is a somewhat thicker layer known as the *pycnocline* (from the Greek *pyknos,* thick, referring to the density of the water), in which the density increases rather abruptly. The increase in density is caused by a cooling of the water, an increase in its salinity, or both.[25] Whichever it is, the change marks the level at which, going downward, wind-caused mixing stops and relative calm prevails. Below the change, in the *deep zone*, the density usually increases very gradually down to the bottom;[26] the average temperature and salinity in the deep zone are 3.5°C and 34.7 parts per thousand.

In the deep zone, currents flow wherever there happen to be horizontal density differences. A density difference arises wherever there is a change in temperature or salinity; this creates a pressure gradient, down which water flows as a gentle current. Such currents are known as *thermohaline currents* (from the Greek *therme,* heat, and *halos,* salt).

Their energy is far less than that of wind-caused currents: their speed is only 1 or 2 km per *day* on average.[27] But they are of great importance as transporters of thermal energy. At some locations, comparatively shallow thermohaline currents descend to become deep currents and then return to the surface; here and there they fork and subsequently rejoin. Taken all together, they form closed loops of enormous extent, which transport thermal energy from ocean to ocean and from the tropics toward the poles. These are thermohaline currents on the grand scale, functioning as energy conveyor belts.

*The Global Energy Conveyor*

The loop current that, with offshoots, spans the whole earth is shown schematically in figure 6.7. It has been dubbed the "global conveyor." The map in the figure combines the representations of several authors and cannot be correct in every detail.[28] This should cause no surprise; the needed data are difficult to collect, and data points are often far apart. The cold, salty currents are at great depth, and the returning warm currents tend to spread out and become less definite as they rise toward the surface. Bear in mind that the conveyor is a three-dimensional structure: the cold currents are close to the ocean floor and ascend slowly to higher levels, warming as they rise. They also vary in salinity. That is, they are thermohaline currents, independent of the winds.

The deep, cold, salty current flowing from north to south for the length of the Atlantic Ocean is the start of the conveyor, insofar as it can be said to have a start. It is believed to come into existence in the following way: The Atlantic is a comparatively narrow ocean. Drying winds from nearby lands cause the

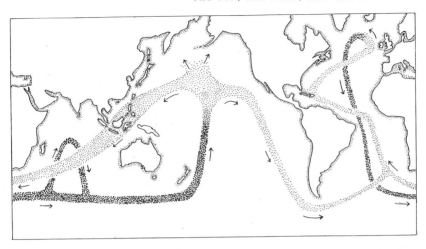

Figure 6.7. The "global conveyor" somewhat idealized. Deep, cold currents are heavily stippled; warmer currents, rising toward the surface, are lightly stippled. Atlantic water is saltier than Pacific water.

ocean to lose more water by evaporation than it gains from rainfall and from inflowing rivers. This makes North Atlantic water much saltier than water at the same latitude in the Pacific.[29] At the same time as it loses heat by evaporation, the water is cooled by cold winds blowing from the Canadian Arctic. The upshot is that  surface water in the vicinity of Iceland becomes steadily colder and saltier—and therefore denser—until it sinks, initiating the conveyor. It flows south and around the tip of Africa.

When it eventually reaches the South Pacific and turns north, the water lost through evaporation in the Atlantic is restored by excess rainfall in the Pacific: the salty water is diluted. Indeed, the conveyor carries salt as well as warmth: it evens out the salinity contrast between the two oceans.

The conveyor's role as a transporter of thermal energy is even more important. In the North Atlantic, before sinking, the conveyor "gives off a staggering amount of heat [to the atmosphere, which] accounts for the surprisingly mild winters of Western Europe."[30] As figure 6.7 shows, in the Atlantic the global conveyor conveys heat from south to north for almost the whole length of the ocean, but in the Pacific the heat flow is poleward both north and south of the equator.

Ocean currents are indeed as important as the winds in carrying warmth from low latitudes to high. But it is difficult to compare the importance of ocean currents and atmospheric winds in achieving the redistribution of energy. According to one estimate, currents are more important than winds in the Northern Hemisphere south of latitude 25° N, whereas at higher latitudes the winds become more important.[31] Depending on the season, however, the surface of the sea is sometimes cooler than the air above it and sometimes warmer, causing warmth to pass repeatedly from air to sea and back again. The atmosphere and the ocean act together in spreading warmth from the tropics to the polar regions.

# 7 | THE ENERGY OF OCEAN WAVES

*Waves of Many Kinds*

At any one moment, the energy in waves in the whole world ocean is only about one-third as great as the energy in currents.[1] All the same, waves are more visible than currents, and they display the ocean's enormous energy much more vividly. Sometimes—as when you watch a stormy sea from a protected shore—they are exhilarating. But when you look up from a small boat at waves rearing over you, they are terrifying.

Waves vary in many ways. They vary in what causes them to form and grow and in what makes them die out and disappear. They obviously vary in size and also in *period*, the time it takes for one wave crest to succeed another at a given point. Taking into account these differences and others, waves can be classified as follows:

*Type 1:* Wind waves (ripples, "ordinary" waves, and swells).
*Type 2:* Internal waves, below the surface of the water.
*Type 3:* Tsunamis, which are usually, but not always, caused by earthquakes.

*Type 4:* Solitary waves, whose solitariness puts them in a class by themselves.
*Type 5:* The tides: to many people's surprise, the rise and fall of a tide is a wave. It could be called, but never is, a "tidal wave," a term often used, incorrectly, for a tsunami.
*Type 6:* Planetary waves: these are not up-and-down waves like all the others. The term refers to the directional changes in a horizontal current that swings alternately to left and right; that is, they resemble Rossby waves in the atmospheric jet streams (see chapter 3).

We consider the first four of these wave types in the following sections, paying particular attention to what causes them (gives them their energy) and how they are dissipated (lose their energy). Type 5, the tides, merit a chapter to themselves (chapter 8). Type 6, planetary waves, are caused by ocean currents responding to the Coriolis effect; though they are technically "waves," they are not waves in the ordinary sense and contain no energy of their own. They are not considered further.

## Wind Waves: Ripples and "Seas"

Wind is the commonest cause of waves. Making and sustaining waves on oceans and lakes is one of the ways winds dispose of their energy. It has been truly said that wave energy is "wind energy that has been temporarily trapped in the waves."[2]

The first question to consider is, How do waves get started? How does the wind exert pressure on calm water? Once the first small ripples have formed, it is easy to see how continued wind pressure will enlarge them. It is not so easy to see how a horizontal water surface is affected by a horizontal wind blowing over it. Nowhere does the wind blow against the water: Why should wind and water do anything more than slide past each other?

This is not a simple problem, and several solutions have been suggested. According to one theory, an apparently smooth sea surface is always dimpled by small variations in air pressure from place to place, and the dimples provide tiny slopes for the wind to act on. A more recent theory proposes that ripples are first caused by downdrafts of wind. The wind is never perfectly horizontal everywhere, and downdrafts produce cat's-paws, the dark patches of ruffled water you see scattered here and there on a calm sea as soon as the air begins to move. The ripples in the cat's-paws provide the slopes that horizontal winds can act on to build up sizable waves.

Ripples and waves don't keep growing: there always comes a time when they disappear and a calm sea is restored. What causes the waves' collapse when the wind dies down and the energy maintaining them stops? Two forces act to flatten an undulating liquid surface: gravity and surface tension. Gravity is by far the more important force, but surface tension alone is enough to even out the smallest undulations. Indeed, this is the technical difference between *ripples* and *waves*. Ripples (this means ripples in a water surface, not ripples in sand) are waves so small that surface tension suffices to flatten them.

It is impossible to specify precisely the maximum size of a ripple—equivalently, the minimum size of a true wave—since it depends on the surface tension of the water, which depends in turn on the water's salinity and temperature. The surface tension of salty water is greater than that of fresh water, and the surface tension of cold water is greater than that of warm water. Ripples usually have wavelengths (the crest-to-crest distance) of less than 2 cm; waves usually have wavelengths of more than 10 cm. A wavelet of intermediate wavelength may be either a ripple or a wave, depending on the surface tension.

Waves larger than ripples are too large for surface tension to flatten; instead, the force of gravity levels the sea surface when the wind stops blowing; gravity is described as the *restoring force*, and the waves are formally called *gravity waves*. Gravity waves (and, a fortiori, ripples) are also leveled by a change of wind: "A sudden reversal of the wind at sea literally knocks the waves flat."[3] (Note that this has nothing to do with the "gravity waves" of modern physics, which are postulated periodic variations in the force of gravity.)

Ripples are of minor importance in the context of energy: their energy is trivial, so in what follows we concentrate on waves. Once they are large enough, waves interact with the wind; the feedback reinforces the waves and makes them grow higher, at the same time increasing the rate at which energy is transferred from the atmosphere to the ocean. So long as the wind continues to blow without abating, the waves grow higher and higher, though not, obviously, without limit. If a wave becomes too steep for the incoming wind energy to sustain it, its crest topples over and it becomes a *whitecap*.

The height to which waves can grow depends on three things: the wind speed; the *fetch*, which is the distance the wind has blown across the ocean without interruption; and the duration, which is the length of time the wind has been blowing with no change of speed or direction. For example if a wind of 5 m s$^{-1}$ (the speed) blows over open sea for 20 km (the fetch), it will take 2.3 hours (the duration) for the waves to grow into a *fully developed sea*, after which they grow no higher; their average height will be about 0.25 m, and there will be a scattering of whitecaps.[4] A wind of 15 m s$^{-1}$, known as a mod-

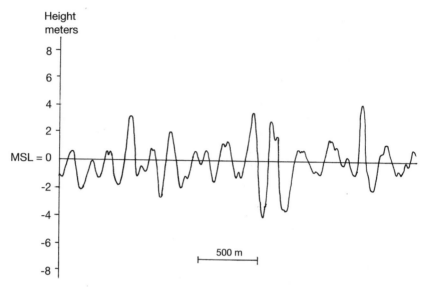

Figure 7.1. Typical profile of a fully developed sea. Note the exaggerated vertical scale. MSL is mean sea level.

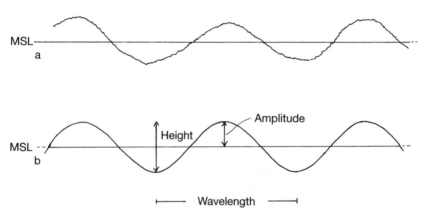

Figure 7.2. (a) Typical profile of swell. (b) A sine wave, showing its dimensions. A wave's height is the vertical distance from crest to trough; its amplitude is half the height. MSL is mean sea level.

erate gale, with a fetch of 480 km, produces a fully developed sea in twenty-two hours. The average height of the waves is about 1 m, most of them are whitecaps, and streaks of foam, or spindrift, begin to blow from their crests. Note that these examples give average wave heights; about one-tenth of the waves will be twice as high or higher. In a fully developed sea resulting from a strong wind, an appreciable fraction of the energy is dissipated by the breaking of the wave crests into whitecaps; then the viscous shearing ("friction") caused by turbulence in the whitecaps produces heat.

The profile of a fully developed sea, indeed of any "sea" in the sailor's sense of the word, meaning a rough sea, is markedly irregular, as figure 7.1 shows. The waves vary tremendously in height, and their wavelengths also vary to some extent, though not nearly so much as their heights. This irregularity makes the physics of real waves much more difficult to investigate than that of the simple, "pure" waves (*sine waves*) shown in figure 7.2. The waves that come closest to sine waves in real life—often they are true sine waves—are *swells*.

## Swell

Wind waves, which in large numbers make a "sea," are found where the wind is blowing. They advance in the direction of the wind and keep on moving. Once they are out of the area where they were generated, they "settle down" (in a manner to be described below) and become *swell*. Figure 7.2a shows the profile of a typical low swell; for comparison a pure sine wave is shown in figure 7.2b, which also shows the terminology used to describe it.

Another note on terminology: the word "wave," by itself, sometimes includes both wind waves and swells; and when there is no risk of misunderstanding it is used as an abbreviation for wind wave.

Swells typically have longer wavelengths and longer periods than wind waves. Few wind waves have wavelengths greater than 130 m, whereas swells are often several hundreds of meters long. Most wind waves have periods in the range 0.2 to 10 s (ripples have periods of less than 0.2 s), whereas most swells have periods in the range 10 to 30 s. These measurements don't define wind waves and swells, however; they are merely typical, and it is quite possible for a wind wave to be longer than 130 m or slower than 10 s and for a swell to be shorter or faster. The defining difference between a wind wave and a swell is origin. A "sea," by definition, is the mass of wind waves in the area where the wind generated them. A swell is the waves (or a single wave, the same word for both) that has traveled outside the generating area.

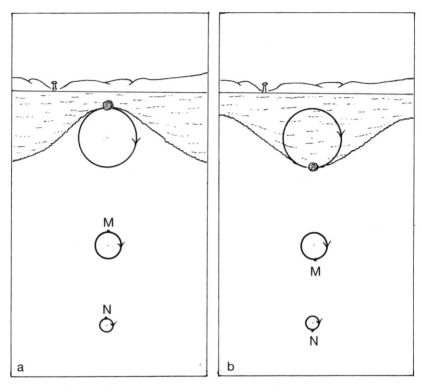

Figure 7.3. Movement of the water in a deepwater wave. The wave is moving from left to right. The two panels are from the same viewpoint (note the identical background scenery). (a) A floating log is on the crest of a wave traveling to the right. (b) Half a period later, the wave has advanced half a wavelength to the right, and the log is in the trough; its orbit is the large circle at the surface. The water molecules M and N, at depth, circle in their own smaller orbits in synchrony with the log.

Before considering the way a "confused sea" (as sailors call it), with high, irregular, foam-tipped waves, becomes converted into a smooth, gentle, regular swell, a digression is necessary on the physics of waves in general.

The first point to emphasize is that when waves travel across the sea, what is traveling is the shape, or *form*, of the wave, not the water itself. This is obvious if you watch an object—a log, say—floating on a choppy sea. Although the waves travel steadily forward, the log does not travel with them; it stays in more or less the same place, bobbing up and down as the crests and troughs of the waves pass beneath it.[5] Therefore the water is not flowing in the direction of the wind, and it obviously is not stationary: How, then, does it move? The answer is shown in figure 7.3. If you could observe the movement of an individual molecule of water, it would be seen to move in circles, as the figure shows. Note that a molecule at the surface—or a floating log—does not go below the surface as it circles, any more than the foot of a person riding a bicycle goes below the pedal: foot and pedal circle together, with the foot attached to the top. In the same way, a water molecule in the sea surface, or a log floating on the sea surface, acts as part of the surface and circles with it.

Water below the surface circles too, in time with the circling water above, but the circles become smaller and smaller at increasing depths; at a depth equal to one wavelength, the circle has a diameter less than one-five hundredth of the diameter at the surface. The speed at which each molecule of water circles is normally less than the speed at which the waveform travels; if it becomes greater, the wave breaks.[6]

Now consider the energy of a wave. Notice first that it has potential energy (PE) because the water is not flat; the PE would disappear if the water were to become flat in response to the restoring force—gravity—acting on it, but so long as it is not flat it has PE. It also has kinetic energy (KE) because of the circling motion of the water in the wave. The total energy of a wave is the sum of its potential and kinetic energies. The PE and the KE of a single wave are equal. As the trough of a wave rises, its KE is converted to PE; then, as the crest sinks, its PE is converted into the KE of the water's circling motion. The latter is dissipated by viscous drag. The energy lost is made good by the wind so long as it is still blowing. If the wind dies down, or if the waves travel out of the windy area, they lose both energy and height and change their form, as described in the next section.

The total energy in sea waves is given by the formula

$$\text{energy} = 1255.68\ H^2 \text{ joules per square meter (J m}^{-2}),$$

in which $H$ represents the height of the waves in meters.[7] The factor 1255.68 is $1/8 \times 1{,}024$ kg m$^{-3}$ (the density of seawater) $\times\ 9.81$ m s$^{-2}$ (the acceleration due to gravity).

The formula gives the energy in joules per square meter of sea surface. It gives more meaningful results with swells than with waves; the waves in a "sea" are very variable in height (see fig. 7.1) and very irregular in shape; swells, on the other hand, are notably uniform in both height and shape, and the shape is often close to a pure sine curve.

Two examples: the energy in waves 1 m high is 1,255.68 J m$^{-2}$. In waves half as high (50 cm) the energy is only one-fourth as much, or 313.92 J m$^{-2}$, because it depends on the square of $H$.

It has been estimated that, at any instant, the energy in the surface waves of the whole world ocean is $10^{18}$ J. When these waves break on shore they release heat—but not much. If all the heat were used to heat the water, with none being lost, it would take 90,000 years to raise the temperature of the world ocean by 1°C. The rate at which wave energy is converted to heat by the waves breaking on all the world's shorelines is believed to be about $2 \times 10^{12}$ J s$^{-1}$ (equivalently, 2 billion kilowatts). The rate at which the sun heats the oceans is 1,500 times as great.[8]

Most of the energy in wind waves and swells is dissipated when they break on the shore. But a fraction is lost while they are still out in the deep ocean—if it were not for this loss, the sea would never be calm.

*How Waves Die Down*

As noted already, the waves in a big "sea" lose much of their energy because of the viscous drag in the turbulent water. What becomes of the remainder? Unsurprisingly, it is also lost because of viscous drag in all the rest of the constantly moving water. What is surprising is the slowness of the loss. The energy captured from a windstorm and carried away from it in smooth swells lasts a long time.

The way waves turn to swells and then fade away is not as simple as it seems. Any train of waves, however jagged its shape, can be analyzed into the sum of a number (usually an infinitely large number) of different component waves, each of them a sine wave like that in figure 7.2b. The component sine waves differ from one another in period, in amplitude, or in position relative to the others—usually in two or all three of these attributes. Figure 7.4 shows a train of waves made up of only three sine waves, as an example (an unnaturally simplified example, to make the figure clear). The heavy line in the upper panel shows the form of the waves to be analyzed.[9] The three sine waves in the lower panel are its component waves.

If the original waves leaving a windy area were to match those shown in the upper panel, their component waves would slowly become separated, as shown in figure 7.5a. The example, as explained, is artificial; in real life the pattern is like that in figure 7.5b: waves of many wavelengths are present, sorted by wavelength with the longest, highest waves in the lead. The sorting hap-

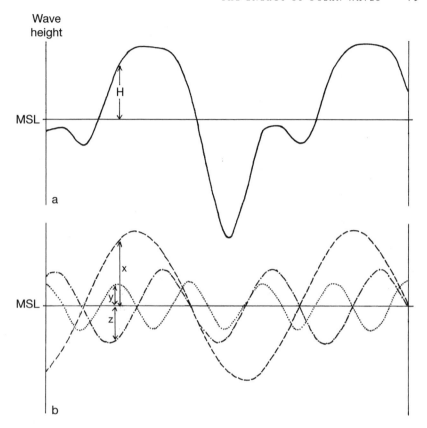

Figure 7.4. Analyzing a train of irregular waves (a) into its component sine waves (b). MSL is mean sea level. The component waves are added as shown. $H$ is the height of the water surface above MSL; $H = x + y - z$, in which $x$, $y$, and $z$ are the heights of the three component waves.

pens because long waves—those with long period and long wavelength—travel faster than short waves.[10]

The sorting process is known as *wave dispersion*. The waves lose energy, and consequently height, because of internal viscous shearing, and the loss happens much faster in short waves than in long; as a result, the waves in the rear shrink faster than those in the vanguard and fade away sooner, leaving the long-wavelength waves as temporary survivors. In theory, a wave with a period of four seconds would have to travel the enormous distance of 23,000

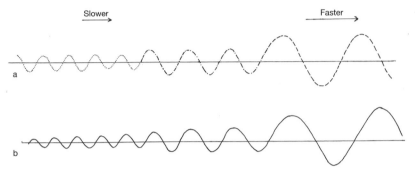

Figure 7.5. The dispersion (sorting out) of waves of different wavelengths; the longer its wavelength, the faster a wave travels. (a) Dispersion of the irregular waves of figure 7.4a. The component waves have separated: the dotted line shows the shortest, slowest waves; the dash-dot line shows intermediate waves; and the dashed line shows the longest, fastest waves. (b) A more lifelike dispersed wave train: the waves vary continuously from long, high, and fast-moving in the lead (right) to shorter, lower, and slower in the rear (left).

km before losing half its height, and the journey would take 1,000 hours (nearly forty-two days).[11] Contrast this with the fate of a wave with a period of one second, which would lose half its height after traveling a mere 12 km, a trip taking 4.3 hours. These numbers illustrate the durability of long waves. Storm waves generated in Antarctica have been observed to travel all the way to the shores of the Alaska panhandle 10,000 km away, arriving as low swells.

Because loss of energy is so exceedingly slow in long swells, the longest ones keep going almost indefinitely; they continue to travel while their height dwindles to a centimeter or two; at this stage their slopes are too gentle for the wind to affect them, and contrary winds cannot stop them. It seems likely that big swells never have enough time or distance to die away entirely. Their end comes when they break on a distant shore.

*At the Beach*

The energy in wind waves and swells is ultimately dissipated when they reach a shore. If they encounter a gently sloping beach, rising at an angle of 10° or less, practically all their energy is spent immediately: the waves break. A wave slows down on moving into shallow water, and its energy is temporarily conserved because it gains height—it rears up. This dooms the wave, however: in rearing up, it becomes so steep that its crest topples forward—it becomes a breaker. This is because the waveform's forward speed has become less than that of the water circling within it. This happens when the wave has grown to a height equal to about one-seventh of its wavelength.[12]

If the slope of the beach is gentle enough, the waves breaking on it lose all their energy. The loss takes place in several ways: some of the energy is converted to turbulence, which abruptly speeds up viscous drag and hence the conversion of energy to entropy; some energy drags beach material—sand and shingle—up the beach and down again; some energizes longshore currents that drag beach material along the shore (see chapter 9); the last of the energy is dissipated as noise—the crash of breakers and the roar of rolling, sliding shingle. The final dissipation of the energy in a big wave is noisy and spectacular.

The course of events just described is not the only possibility. If waves reach a steep shoreline, they are partly or wholly reflected back to sea, where they interact with incoming waves to produce confused choppiness. Confused choppiness—a "sloppy sea"—is a sign that viscous shearing is proceeding vigorously and wave energy is rapidly being dissipated. Thus, one way or another, wave energy inevitably stops at the beach.

## Internal Waves

The waves we have considered so far have all been surface waves; the movement of the water dies away a short distance below the surface, becoming negligible about one wavelength down. But that is not to say there is no wave action at greater depths: on the contrary, there are often big *internal waves* under the surface; they are invisible, as waves, from above. They can be detected, however, if they are not too deep and you know what to look for, because they produce surface *slicks*.

A slick is a band of smooth water forming a lane across a gently ruffled sea. Several widely spaced slicks can often be seen, all somewhat curved and more or less parallel. They appear because currents flow from the surface downward into the troughs of the internal waves from either side (fig 7.6). If there is an oily surface film produced by ships and boats, or naturally by living organisms in the water, the film thickens where the currents converge. The thickened oil film makes the water appear glassy—hence the slicks, which show best when the water between them is slightly ruffled. Slicks are a common part of the scenery for ocean watchers, scenery made more interesting if you visualize the unseen internal waves below.

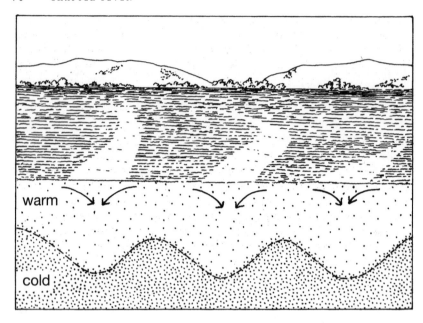

Figure 7.6. Internal waves, with slicks at the surface above the wave troughs. From E. C. Pielou, *Fresh Water* (Chicago: University of Chicago Press, 1998).

Internal waves form most readily in the shallow waters over coastal shelves.[13] They develop where rising and falling tides are channeled by seafloor valleys, becoming concentrated into currents. Where these currents—they could be described as seafloor rivers and streams—flow over topographic irregularities, they develop waves, just as a river on land does when it flows over shallow rocks. Among the causes of internal waves far out to sea are moving low-pressure systems that produce the inverted barometer effect (see chapter 6); quickly repeated changes in wind stress also cause them.

Internal waves are undulations in an *internal surface* in the sea, just as surface waves are undulations in the "ordinary" surface. The internal surface is the layer known as the pycnocline (see chapter 6), in which the density of the water increases relatively suddenly as you descend from the surface, either because of cooling or because of increased salinity. The more sudden the density change, the more sharply defined the internal surface is, though it is never infinitesimally thin, as the air-sea surface is. When the density change is comparatively gradual, the layer in which it takes place hardly merits the name surface; in any case, internal waves can develop whether the density change is abrupt or gradual.

In what follows, we assume the internal surface is well defined. The difference in density between the waters above and below it are orders of magnitude less than the difference in density across an air-sea interface: this is what accounts for the striking dissimilarity between internal and surface waves. Compared with surface waves, internal waves are much higher, have much longer periods, and move much more slowly. These characteristics are most pronounced in the deep ocean, where the density contrasts are even less than they are close to the shore. Internal waves 200 m and more high, with periods of several hours, have been recorded in the open ocean.

An internal wave has much less energy than a surface wave of equal height because the difference in density is so slight. The energy in a surface wave 1 m high is, as we saw earlier, 1,256 J $m^{-2}$; the energy in a 1 m high internal wave in the open ocean is less than 4 J $m^{-2}$. The internal wave would have to be 18 m high to have the same energy as the 1 m surface wave.

Internal waves dissipate their energy more quickly than surface waves do and cannot travel nearly as far before dying away. They break when they run up a sloped seafloor in the same way that surface waves break when they run up a sloping beach; internal waves even produce "internal surf."[14] And although the energy in underwater breakers is slight compared with the energy in subaerial breakers, it does biologically useful work nevertheless; it is the energy that, by causing turbulent mixing, prevents the water close to the seafloor from stagnating.[15]

*Tsunamis*

A *tsunami* is a group of enormous waves set in motion by a sudden disturbance on the seafloor. The disturbance is usually an earthquake, but other causes are possible, such as a submarine landslide, a slumping of the seafloor, or a submarine volcano. Probably tsunamis are often caused by an earthquake plus a submarine landslide triggered by it. A meteorite falling into the ocean can cause a tsunami, as can a vast rockfall from cliffs bordering the sea; these last two causes do not, of course, originate on the seafloor, so perhaps a tsunami could be more exactly defined—though it never is—as a group of enormous waves caused by a sudden, unpredictable, short-lived natural calamity.

It is also necessary to say what a tsunami is not. It is *not* a tidal wave; tsunamis have nothing whatever to do with the tides. The misuse of the term "tidal wave" to mean a tsunami may have arisen because a tsunami wave sometimes looks like an unusually high tide that has risen exceptionally fast at an unexpected time; primitive people seeing a tsunami for the first time would have been puzzled and may have misidentified the cause.

The most noteworthy facts about tsunamis are, first, that they are nearly always caused by earthquakes and therefore derive their energy from the earth's internal energy, and second, their awe-inspiring size when they reach shore. Presumably little tsunamis, triggered by small earthquakes or landslides, are common, but they go undetected because they are masked by the sea's continual movement. A tsunami has to be big to be recognized; a group of waves caused by an earthquake weaker than about 6.5 on the Richter scale could go unnoticed. Big tsunamis are the product of truly energetic earthquakes: it has been conjectured that only 1 percent of a submarine earthquake's energy becomes converted to wave energy.[16]

Although the waves of a typical tsunami are "big," this does not mean they are high before they reach shore; out in the open ocean far from land they are seldom more than a meter high, and they attract no attention from the passengers and crew of a ship whose path they cross. In the open ocean their bigness consists in their exceedingly long wavelengths and periods compared with those of swells. Tsunamis have wavelengths of hundreds, sometimes thousands, of kilometers and periods several hours long; their energy is therefore thinly spread until they reach shore; but when they do come ashore, they are brought to an abrupt stop from a speed that may exceed 900 km/h: the collision turns an unremarkable wave into a killer.

About two-thirds of all tsunamis happen in the Pacific Ocean, because earthquakes are so numerous in the "ring of fire" around the Pacific. The earthquakes most often responsible are those caused by tectonic plates grinding against each other, as one plate is subducted under another (see chapter 15). Tsunamis often travel huge distances; for example, the tsunami generated by the 1960 earthquake off the coast of Chile traveled across the whole width of the Pacific to Japan; the distance was 17,000 km, the time taken twenty-two hours, and the average speed 773 km/h.[17]

In spite of crossing the Pacific, this tsunami, like all tsunamis, was technically speaking a *shallow-water wave*. The term applies to any wave having a wavelength greater than twenty times the depth of the water. The average depth of the ocean is less than 4 km, and the wavelengths of tsunamis are always hundreds of kilometers, which means they all rank as shallow-water waves. The defining characteristic of a shallow-water wave is that its speed is governed by the depth of the water. The wave is said to "feel the bottom," and it experiences appreciable drag all along its path. The speed of a shallow-water wave is given by the formula

$$C = \sqrt{(gd)} \text{ m s}^{-1} \text{ or, equivalently, } 3.6 \times \sqrt{(gd)} \text{ km/h};$$

here $C$ is the wave's speed, $d$ is the depth of the water in meters, and $g$ is 9.81 m s$^{-2}$, the acceleration caused by gravity. A tsunami traveling through water 500 m deep, for example, will have a speed of about $3.6 \times \sqrt{(9.81 \times 500)}$ km/h = 252 km/h.

The waves of a tsunami change their speed as they travel, slowing down where the water becomes shallower and speeding up if it deepens again; the separate waves of a single tsunami don't maintain the same speed: each goes at the speed appropriate to the depth of the water below it, which is continually changing as the wave travels over the surface above an uneven ocean bottom.

Tsunamis dissipate energy while they travel, but in a somewhat different way than swells do. Two of the differences are noteworthy.

First, because of their enormous wavelengths, tsunamis experience the drag of the seafloor wherever they are, whereas swells feel the bottom only when they are comparatively close to shore. Figure 7.7 shows why this is so; it illustrates how the water moves within a shallow-water wave and should be compared with figure 7.3, which shows a deepwater wave. In the deepwater wave, the circular orbits of individual molecules of water shrink to nothing at some distance above the seafloor. By contrast, in a shallow-water wave the water moves in elliptical orbits that get progressively flatter the greater the depth, with the water at the bottom simply moving back and forth. The back-and-forth swishing drags constantly on the seafloor, dissipating the wave's energy.

The second difference in the way tsunamis and swells dissipate their energy is this. Tsunamis are generated by the jolt of an earthquake at a single spot on the ocean floor, and the resultant waves spread out in expanding circles; apart from the difference in scale, they resemble the rings of waves spreading outward when a rock is dropped in a pond. This means that the energy is spread along the circumference of an ever expanding circle; the total amount of energy is unaltered, but it becomes more and more thinly spread.[18]

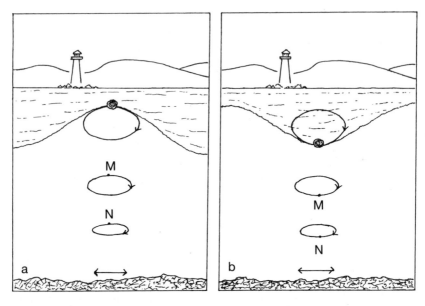

Figure 7.7. Movement, from left to right, of the water in a shallow-water wave. This figure shows the wave in figure 7.3 closer to the shore, where the water is shallow. Note the elliptical orbits of individual water molecules narrowing at progressively greater depths. At the seafloor, the water shifts back and forth in a horizontal plane.

By contrast, swells spread much less (fig. 7.8). As they leave the storm area where their parent wind waves were generated, they all advance parallel with the wind; once they are beyond the wind, they diverge to some extent, but rarely more than 45° to left and right of their original direction.[19]

In spite of spreading and "friction" with the bottom, powerful tsunamis often wreak enormous damage when they reach the land. Huge waves surge up the shore with tremendous force and inundate low-lying coastland; individual waves come at intervals that are sometimes an hour long. The higher the tide at the time each wave arrives, the farther inland it can go, destroying much that lies in its path. The tsunami in Chile in 1960 destroyed villages along an 800 km stretch of coastline before traveling to Japan, where it was still able to do much damage. The 1964 tsunami originating near Anchorage, Alaska, killed 107 people there before traveling outward over the Pacific and southward along the North American coastline; it killed 61 people in Hawaii and 12 in Crescent City, California.

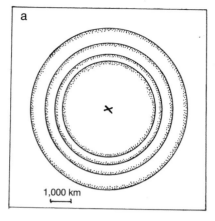 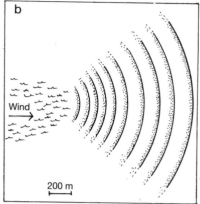

Figure 7.8. The spreading waves of (a) a tsunami and (b) a swell. Note the different scales of the two maps. In (a) the X is over the site of the earthquake that caused the tsunami, which is a group of four waves; the wave with the longest wavelength spreads the fastest. In (b) the choppy patch of sea is the location of the storm where the wind waves giving rise to the swell originated.

Not all killer tsunamis come from afar. If an earthquake, even a compara-tively minor one, shakes the seafloor where thick accumulations of sediment are poised ready to slump down the continental slope (the steep submarine slope between the continental shelf and the deep sea), the disturbance is likely to cause a disastrous tsunami on the nearby coast. Probable examples caused in this way are a 1992 tsunami on the shores of Nicaragua that killed 170 peo-ple; a 1993 tsunami on the shore of Okushiri Island, Japan, that killed more than 200; and the catastrophic tsunami that struck Papua New Guinea in 1998, sweeping away several villages and killing more than 2,500.

## Solitary Waves

The waves we have considered so far have all been *periodic,* that is, repetitive, in the sense that one wave follows another, on and on and on, theoretically ad infinitum. Solitary waves, by contrast, occur singly, as their name implies. If that were all, there would be little more to say. But it is not all. A solitary wave is so unlike a periodic wave that it seems inappropriate to call it a "wave." In-deed, because of their extraordinary behavior, another name has been devised: they are now sometimes called *solitons,* a name used most often for solitary electromagnetic waves (see chapter 18). Before describing their weird behavior, it is worth giving an outline of their discovery.

The first person known to have seen and described a solitary wave was the Scottish naval architect John Scott Russell.[20] He described what he saw to a meeting of the British Association for the Advancement of Science in 1844. He had been watching a horse-drawn boat being towed along a narrow canal when "the boat suddenly stopped—not so the mass of water in the channel which it had put in motion; it accumulated round the prow of the vessel in a state of violent agitation, then suddenly leaving it behind, rolled forward with great velocity, assuming the form of a large solitary elevation, a rounded, smooth and well-defined heap of water, which continued its course along the channel apparently without change of form or diminution of speed." Russell goes on to say that he followed the wave on horseback as it traveled at eight or nine miles an hour. In form it was "some thirty feet long and a foot to a foot and a half in height. Its height gradually diminished, and after a chase of one or two miles I lost it in the windings of the channel."

This first reported example of a soliton was in a channel. Solitary waves have more recently been discovered in the open ocean, specifically in the Andaman Sea near Thailand.[21] They also occur, invisibly, in the atmosphere.[22] Their salient characteristic is that, unlike periodic waves, they do *not* disperse. Rather than changing into a dwindling series of smaller waves as each member of a group of periodic waves does (see fig. 7.5), a solitary wave, once started, becomes more and more distinct—steeper and higher. It proceeds in solitary state at undiminished speed.

The energy in solitary waves is considerable. The solitary waves in the Andaman Sea (which were solitary internal waves) were strong enough to push an oil rig nearly 30 m and spin it through a right angle; solitary waves in the atmosphere have been found to cause a rise in air pressure, gusty winds, and rain. These are occasions when solitary waves are seen to dissipate energy by doing work, in the scientific sense. Their energy is also dissipated, slowly but inevitably, by viscous shearing and conversion to entropy. If it were not for viscous shearing, they would remain unchanged forever, because, as already noted, they do not disperse.

This amounts to saying that the laws of physics describing ordinary periodic waves do not apply to solitary waves. The two kinds of waves differ fundamentally: solitary waves have their own physical laws, and unraveling them is now a fast-growing branch of science.[23]

 # 8 THE ENERGY OF THE TIDES

## Tides as Waves

The movements of the ocean are traditionally classified into currents, waves, and tides. In one sense, however, tides *are* waves, albeit of a special kind. Unlike wind waves and swells, which are caused by the wind, and tsunamis, which are caused by earthquakes and similar disturbances, tides are caused by the gravitational pull of the moon and sun. They are waves nevertheless; to distinguish them from other waves, we shall call them tide waves.

A *tide wave*, as we noted in chapter 7, should not be confused with a "tidal wave"; the latter is a misnomer for a tsunami. Tide waves have some very special properties. They are controlled predominantly by the moon—the effect of the sun is only half as great—and because of this their period is equal to one-half a lunar day on average, or twelve hours and twenty-five minutes (a lunar day, twenty-four hours and fifty minutes, is the time it takes for the earth to rotate so that the moon makes one complete circuit, from a given compass direction on one day to the same compass direction on the following day). Because the lunar day is fifty minutes longer than the twenty-four-hour solar day, the tides come roughly fifty minutes later by the clock on each succeeding day. The wavelength of a tide wave is about 22,000 km, which is half the earth's circumference at the equator.

The reason these figures are not exact is that the tides are affected by the sun as well as by the moon, and the relative positions of sun and moon are always changing. Moreover, a tide wave is a shallow-water wave because its wavelength is so great relative to the depth of the ocean that its speed varies from place to place as it travels. It slows down if the water becomes shallow and speeds up again where it deepens. Because of this, high tide comes late wherever a shallow continental shelf extends a long way to seaward of the shoreline.

The range of the tide (the difference in water level between high and low tide) also varies from place to place. The range depends on the shape of the shoreline and the pattern of the depth contours: it is often especially great in deep, narrow inlets where an entering tide wave piles up and has nowhere to spread out.

Normally there are two high tides and two low tides per lunar day; that is, the tides are *semidiurnal*. The highs are not equally high, nor are the lows equally low. The relative heights vary from tide to tide and from day to day, depending on the distances of the moon and the sun to the north and to the south of the equator, which vary from day to day all through the year. As shore dwellers know well, at most places the sequence of tides in a lunar day is higher high, higher low, lower high, and lower low. But in a few places, for a few days in each lunar month, the tides are *diurnal*, with only one high tide and one low tide per lunar day. This is most surprising when you consider how the tides are caused, as we do in the next section. Diurnal tides happen when the low tide between two high tides is itself so high that the high tides before and after it seem to be one long, uninterrupted high tide; likewise, two succeeding low tides seem to be one long low tide. Two further complications deserve mention.

First, there are internal tide waves, in other words, *internal tides*. These are tides affecting an internal surface in the sea, the surface whose undulations are the internal waves described in chapter 7.

Second, there are *tide currents* or *tide streams*.[1] These are the currents that flow back and forth as the tide rises and falls. Along a coastline with numerous offshore islands separated by winding channels, tide currents are forced to follow correspondingly winding courses. At each turn of the tide, the current reverses direction; its speed is greatest at midtide. The unique characteristic of tide currents is that, in contrast with all other currents, the speed of flow is the same through the whole depth of the water, from the surface right down to the level, close to the bottom, where drag slows it.[2] In the deep ocean, the tide current is slow: the water shifts about 1 km during each half-period (of 6 h, 12.5 min) so the average speed is only 160 meters an hour. In shallow inshore waters, a tide current flows much farther between each reversal of direction, at speeds that may exceed 1 km/h.

The movement of the water molecules within a tide wave is an exaggerated form of the movement shown in figure 7.7 for a shallow-water swell. The wavelength of the tide wave is more than five thousand times the water's depth; consequently the elliptical paths of the molecules are so flattened as to be indistinguishable from horizontal straight lines; figure 7.7 can be modified to represent the tide wave by replacing the stack of ellipses with a stack of double-headed horizontal arrows all of the same length.

If you live near a gently sloping ocean beach, you can watch these currents any time the water is glassy calm: what you see is the water creeping slowly landward up the beach as the tide rises and then down the beach as it falls, back and forth twice a day, endlessly.

That each tide is a "wave" and causes "currents" should not obscure the fact that tide waves and tide currents are utterly different from other waves and currents, because their causes are extraterrestrial. The causes and consequences of the tides are always in step over the whole world; in contrast, the causes and consequences of all other ocean movements are always regional or local.

### The Energy That Drives the Tides

It is often said (for example, in the preceding section) that the tides are caused by the gravitational pull of the moon and the sun on the oceans. This is true, but it is so oversimplified as to disguise what's really happening.

Disregard the sun for the moment—it is much less important than the moon in the context of tides—and visualize the *earth-moon system*. The earth has a mass of $6 \times 10^{24}$ kg (or more impressively, 6 billion trillion metric tons), eighty-one times that of the moon (which, at more than 70 billion billion metric tons, isn't negligible). The two bodies are about 384,400 km apart.[3] Each exerts a gravitational pull on the other, so why don't they fall in on each other and become one?

The answer is that the system is rotating, like a barbell with very unequal weights, around an axis through its center of gravity (CG). Figure 8.1 illus-

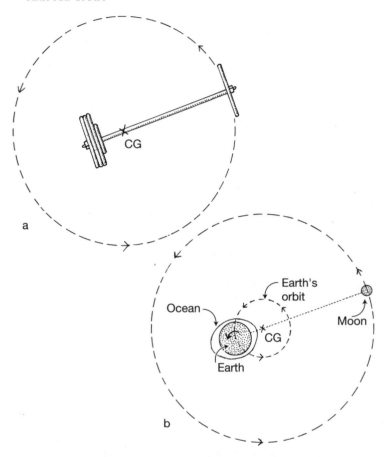

Figure 8.1 (a) A lopsided barbell rotating counterclockwise around its center of gravity (CG). (b) Analogous rotation of the earth-moon system, as seen looking down on the North Pole. Not to scale: the moon is disproportionately large relative to the earth, which brings the system's CG outside the earth instead of inside. The whole earth-moon system rotates once in 27.32 days; the earth also rotates around its own axis once in twenty-four hours relative to the sun.

trates the resemblance except that it shows both CGs located somewhere between the two masses—weights in the case of the barbell, heavenly bodies in the case of the earth-moon system. In reality, because the mass of the earth is so much greater than that of the moon, their CG is inside the earth; it is below the ground, about one-fourth of the way from the surface to the center. In the figure, the CG has been placed in the space between the two bodies merely for clarity and to emphasize that the moon's orbit is centered *not* on the earth's center, but rather on the CG of the earth-moon system. That it is inside the earth makes no difference to the dynamics.

Because the whole system is spinning, at the rather stately speed of one revolution each 27.32 days, the two bodies are kept apart by centrifugal force.[4] The distance between them and the speed of rotation of the system are in equilibrium: that is, the gravitational attraction tending to make the bodies fall together exactly balances the centrifugal force tending to drive them apart. Which raises the next question: Why is the system spinning?

Think of a child's top. If you put it on the floor, it just lies there; it will spin only if you give it a sharp twist to impart *rotational energy* to it. Likewise with the earth-moon system; it has rotational energy. Where does the rotational energy come from? Undoubtedly it is part of the original rotational energy of the solar nebula—a vast rotating cloud of dust and gas—that was the precursor of the solar system.[5] As the primordial dust gradually accreted into solid bodies and groups of bodies, they retained their rotational energy, and much of it (not all, as we shall see) persists. In a nutshell, the energy that drives the tides is a fragment of the rotational energy of the infant solar system, surviving in the rotation of the earth-moon system.

Figure 8.1 also shows what happens to the ocean; note that for the present we are still disregarding the presence of the sun, and also of the continents, which cover less than 30 percent of the earth's surface. Because the earth-moon system spins, the shell of ocean water encasing the earth is deformed: its surface takes on the shape of a football, with one end pointing toward the moon, the other end away from it. In the figure the football's length is greatly exaggerated, for clarity; the difference in depth between the deepest water and the shallowest is really only a meter or two. The deformation arises because the moon's gravitational pull raises the water nearest it up into a bulge, while at the same time the water on the side farthest from the moon, and therefore on the outside of the rotating system, bulges because of centrifugal force.

More precisely, gravity acts to pull both earth and ocean toward the moon, while centrifugal force drives both earth and ocean away from the moon. The effects balance: on the side of the earth nearest the moon, the moon's gravitational pull exceeds the centrifugal push, and vice versa on the side farthest from the moon; hence the symmetry of the football.

Now recall that the earth rotates on its axis, relative to the moon, once per lunar day of 24 h, 50 min. As it does so, every point on the surface passes, suc-

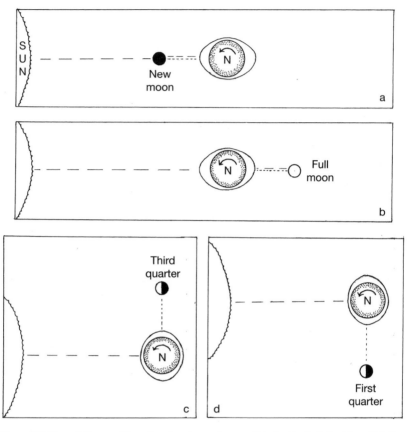

Figure 8.2. The relative position of earth, sun, and moon; N is the North Pole. (a) and (b) The three bodies are aligned, giving spring tides. (c) and (d) The three bodies form a right angle, giving neap tides. Note that the outline of the ocean forms a more elongated ellipse at spring tides than at neap tides, giving a bigger tide range. (Both c and d are viewed from the side; the sunlit half of the moon is white. Seen from the earth, the first-quarter moon appears D-shaped, and the third-quarter moon appears C-shaped.)

cessively, first under an ocean bulge, then under a shallow part, then another bulge, and then another shallow part of the "football." This represents the familiar sequence of high tide, low tide, high tide, low tide.

Next we consider two details that complicate this simple picture. The first detail is the sun. In the mathematical theory of the tides the sun's effect is ex-

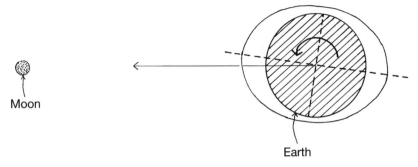

Figure 8.3. The axis through the tide waves (the ocean "bulges" raised by tidal forces) does not coincide with the line joining earth and moon; it lags behind because of drag.

ceedingly complex, but all we need to know here is that the sun reinforces the effect of the moon when earth, sun, and moon are all aligned (as they are at new moon and full moon) and partly negates its effect when the sun's gravitational pull is at right angles to the moon's (as it is when the moon is in its first and third quarters). Figure 8.2 shows what happens. In each month, the tide range reaches a maximum around the time of new moon and full moon, giving *spring tides;* tide range is at a minimum around the times of first quarter and third quarter, giving *neap tides.*

The second complication is illustrated in figure 8.3. It is crucial from the energy point of view. The figure shows, more precisely than was possible in small-scale diagrams, the orientation of the football-shaped shell of ocean encasing the earth. Notice that its long axis is not along the line joining earth and moon; it is deflected through a small angle. This is because the crests of the two tide waves (the bulges at the ends of the football) cannot keep up with the moon: they lag behind, held back by drag between the water and the ocean floor. The delay at any observing station on the coast depends on the depth of the water offshore and on the topography of the bottom. If the water is deep and the bottom flat, the lag will be short, but if the water is shallow and the bottom hilly—perhaps with islands breaking the surface—the lag may be several hours. The lag shows that drag is operating and, consequently, that energy is being dissipated.

*The Dissipation of Tidal Energy*

The energy of the tides is dissipated by the drag of water on the ocean floor and along the shorelines of the world; the drag is especially strong in shallow seas overlying continental shelves. As always, drag produces heat, increasing the entropy of the oceans. The rate at which heat is produced is believed to be about $2.5 \times 10^{12}$ W—think of the energy from 25 billion 100 watt lightbulbs.[6]

Drag also causes the earth to lose rotational energy, a loss that is going on all the time. The drag of the tides slows the rotation of earth so that, imperceptibly, the days become longer and longer. Here we are considering the rotational energy not of the earth-moon system as a whole, but of the earth as a single body, spinning on its axis once every twenty-four hours relative to the sun.

Now for some numbers. What is the rotational energy of the earth (also called, more briefly, the *spin energy*), and how is it computed?

Recall (see chapter 3) that any object moving in a straight line has kinetic energy, KE. Further, if the mass of the object is $M$ kg and it is moving with velocity $v$ meters per second, then we can compute its KE from the formula KE $= \frac{1}{2} Mv^2$ J. Likewise, a spinning object, such as the earth rotating on its axis, has energy by virtue of its spin, but a different formula is required to measure it, a formula using the object's *moment of inertia, I,* and its *angular velocity, w.* The formula is spin energy $= \frac{1}{2} Iw^2$ J; it is of the same form as the formula for KE, so that if we know the relevant values of $I$ and $w$, the spin energy is easy to compute. Before doing so, however, it is necessary to explain moment of inertia and angular velocity.

First, moment of inertia, $I$: in the same way that a nonrotating body's mass is a measure of its resistance to being "pushed" (accelerated or decelerated) in a straight line, a rotating body's moment of inertia is a measure of its resistance to having its rate of spin altered (either increased or decreased). The moment of inertia depends on the mass and *also* on the shape of the body and the location of its spin axis. For example, imagine two gates of the same weight, one wide and one narrow; it is obvious from experience that the wide gate will be harder than the narrow gate to swing on its hinges. This is equivalent to saying that it has a greater moment of inertia. Similarly, imagine two flywheels of the same weight, one made of aluminum and the other of lead, and suppose they are spinning at the same speed; it will take more effort to slow the aluminum wheel than the lead one, because its diameter is greater, and therefore its moment of inertia is greater. Without going into details on how the moment of inertia of a body is computed, it suffices to say that the earth's is[7]

$$I = 8.04 \times 10^{37} \text{ kg m}^2.$$

Next consider angular velocity, $w$: this is the rate at which a body spins, measuring the angle in radians (1 radian $= 57.3°$). For the earth,

$w = 1$ rotation per 24 hours $= 7.27 \times 10^{-5}$ radians per second.

Then, for the spin energy of the earth, we have

$$\tfrac{1}{2}\,Iw^2 = 2.125 \times 10^{29}\ \text{J}.$$

This is the spin energy at the present time. It is being gradually lost because of the braking effect of the tides, which slows the earth's rotation. The days are lengthening at the rate of 0.0024 seconds per century.[8] If the rate continues, the day will be ten minutes longer than at present 25 million years hence, 20 minutes longer 50 million years hence, and so on.

This "lost " rotational energy is not truly lost, nor is it converted into entropy. On the contrary, it is conserved: as the earth slows down, energy is transferred outside the earth to the earth-moon system as a whole. The rotation rate of the system increases, causing the distance between earth and moon to increase too. In brief, the rotational energy that originally belonged to the earth alone is gradually being shared with the earth-moon system.

## Other Tides

Tides affect the atmosphere and also the "solid" earth itself. As you would expect, the atmosphere is as strongly affected as the ocean by tidal forces, and at first thought it is surprising that atmospheric tides should be so elusive—nobody notices them. The atmosphere does indeed respond to tidal forces as it should, but the effect is almost completely masked by volume changes resulting from the diurnal heating and cooling of the atmosphere by the sun. As a result, the atmospheric tides are imperceptible without sensitive instruments. Their contribution to the earth's energy balance is negligible compared with that of ocean tides.

The great contrast between the atmosphere and the oceans in the prominence of their tides arises chiefly because air is compressible and water (almost) incompressible. Water's expansion and contraction in response to temperature changes are much too slight to mask the tides.

The solid earth is slightly affected by tidal forces too, because it is not rigid. The shape of the earth is deformed in the same way that the ocean is deformed, but only to a minuscule extent: the crustal bulges facing toward and away from the moon are less than a meter high on either side of the earth, a sphere 12,750 km in diameter.

# 9 HOW SURFACE ENERGY SHAPES THE LAND

*Sources of the Energy*

Land surfaces everywhere are nearly always uneven or hilly to some extent, even where there are no mountains; expanses of truly flat land, such as dry lake beds, are always surrounded by higher ground. Wherever you look, the land has *relief* or "non-flatness."

This statement is so obvious that it goes without saying, and like many other such statements, it deserves more attention than it usually gets. Why is it true, and what are its implications? Put briefly, the answer to the first question is that land is raised into hills, ridges, mountains, and volcanoes by the earth's internal energy (as described later, in chapter 15). The raised surfaces are simultaneously worn down by wasting and erosion, which sometimes smooths the relief and sometimes—as when rivers erode deep valleys, for instance—exaggerates it. The energy of these external agents is the topic of this chapter.

Now for the implications of the fact that the land isn't level everywhere. Energy from the earth's interior, in deforming the earth's surface and lifting up mountains, imparts to it potential energy—specifically, gravitational PE (see chapter 2). Recall that gravitational PE is always relative to some chosen base level, usually sea level. This means that any chunk of rock or soil above sea level will move downward if something happens to dislodge it. If it cannot do so now because it is already at the bottom of a valley or hollow, it may be able to in the future when the object's surroundings have been eroded away, leaving it "poised" for a fall. In the few places on earth where the surface is at present below sea level, for example, the valley of the Dead Sea and Death Valley in California, the surrounding slopes have PE relative to the respective valley bottoms.

We now ask how the potential energy of the land is liberated. Or, which comes to the same thing, how the earth's crustal material shifts from higher to lower elevations. Two processes are involved: *mass wasting* and *erosion*.

Mass wasting is a general term that includes landslides, rockfalls, earth flows, earth slumps, debris flows, mudflows, and soil creep.[1] The downhill movement of snowbanks and sloping snowfields in mountainous country is another, often disregarded, form of mass wasting: everything from sudden avalanches to the slumping of crusted snow to the slow downhill dribbling of "crumbs" of snow is included here. By definition, mass wasting is the downward movement of material caused solely by the pull of gravity.

Erosion, the other process that shifts earth materials from high elevations to low, is transport by flowing water, wind, or glaciers. The need for a medium of transport distinguishes erosion from mass wasting. The two processes resemble each other, however, in that both act on separated chunks of material, sometimes big blocks of rock in the case of mass wasting, more often tiny particles of sand, silt, and clay—"grains" rather than "chunks"—in the case of erosion.

This raises the problem of what breaks rocks from the earth's solid bedrock in the first place and then fractures the detached blocks into smaller and smaller particles. The process is called *weathering*.

## Weathering

Weathering is the disintegration of surface rock. Wherever bare rock is exposed at the surface, it is always cracked or broken to some extent, sometimes into giant blocks, sometimes into an expanse of sharp-edged rock fragments, sometimes into a layer of crumbs or flakes. The inorganic ingredients of the soil, everything from coarse sand to minute clay particles, are ultimately derived from bedrock by various weathering processes.

Before we go into the details of how rocks disintegrate, it is worth asking what holds them together in the first place. A more inclusive question is, Why doesn't any solid object, be it a rock or a teacup, fall to pieces spontaneously? The object must consist of a collection of atoms and molecules, so why don't the atoms and molecules simply lie there like a pile of dust? What holds them together in definite, recognizable shapes?

These questions are the subject matter of solid-state physics; as with all branches of science, the more you know, the more aware you become that unlimited fields of discovery lie ahead. All we need say here is that what makes solid objects solid is chemical bonds (about which more in chapter 10) and that breaking anything solid, rocks included, entails the rupture of chemical bonds, a process that consumes energy.

Now, briefly, for the details. Rocks disintegrate in two ways: by *mechanical weathering* and by *chemical weathering*. In spite of the names, both kinds of weathering entail the breaking of chemical bonds, but only chemical weathering involves chemical reactions in the ordinary sense.

Mechanical weathering at the surface acts on rocks already cracked while they were deep underground. Igneous rocks form when magma (the hot, molten rock at depth) cools and solidifies; the magma shrinks as it crystallizes, and the shrinking produces fissures. Sedimentary rocks, formed when loose sediments become cemented, are also apt to crack; tremendous pressures develop when tectonic plates move against each other, forcing sedimentary strata to bend and fold. This sets up tensions that cause intermolecular chemical bonds to break and fine fissures to develop on the outer sides of the folds.[2]

Then erosion removes the tremendous weight of material lying on top of the rocks; although they have cracked while deep underground, the fissures have been kept tightly closed by the pressure of the surrounding material. Removal of the overlying sediments allows the rocks to expand: molecular bonds that had been compressed and shortened now lengthen, and they snap if they are stretched too far.[3] Some of the fissures enlarge into clean breaks. The whole process is called *unloading*. Once the fissured rocks are exposed to the air, true mechanical weathering can begin (fig. 9.1).

It happens in a variety of ways, one of which is *thermal cracking*. When rocks are heated by the sun on a sunny day they expand; then as they cool by radiating heat into a clear sky at night, they contract. The alternate expansion and contraction cause further fracturing.

In cold climates, thermal cracking is augmented by *frost cracking:* water penetrates exposed fissures and freezes when the temperature falls below the

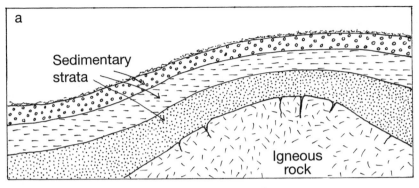

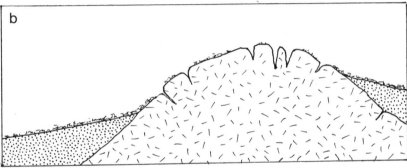

Figure 9.1. How mechanical weathering starts. (a) Magma from the depths has welled up to form a dome of igneous rock below sedimentary strata; fine cracks have formed as the magma crystallized. (b) Millions of years later. Most of the sedimentary strata have been eroded away; relieved of pressure, the cracks have widened, especially those in the outcrop with no load on them. Frost cracking widens them further.

freezing point. Water in pores in the walls of the fissures also freezes, as does water in pores deeper in the rock, which migrates to the newly formed ice and freezes onto it. The result is an increase in the volume of water trapped in a fissure and freezing there. It expands as it freezes until it cracks the rock.[4] Water expands on freezing because the geometrically arranged water molecules in ice crystals occupy more space than they did while the water was liquid.[5]

Let's consider the energy exchanges in frost cracking. As the water filling a fissure cools, it loses some of its thermal energy by radiation. But not all: the rest is stored as chemical potential energy in the molecular bonds in the ice crystals. This PE is "spent" in stretching and finally rupturing the chemical bonds of the rock. It's worth repeating that when a solid object breaks, what breaks are the chemical bonds holding it together; that's what breaking *is*.[6]

Not all mechanical weathering begins with the cracking of bedrock, however. Rocks sometimes grind together with enough force to pulverize each other, yielding fine-grained *rock flour*. The "rock milling" happens when boulders embedded in the base of a glacier grind against the rocks below. Rock are also crushed and pulverized along geological fault planes.

Now for chemical weathering: it is the disintegration of rock as a result of chemical reactions. Every chemical reaction entails an energy change, sometimes a gain, sometimes a loss. In weathering reactions the change is always a loss: energy is liberated.

In the chemical weathering of rocks, the most frequent reactions are those caused by weak acids attacking and dissolving some of the rocks' component minerals. Weak acids are much commoner than pure water in the natural world. Rain, for example, is always slightly acidic because as it falls it dissolves a fraction of the carbon dioxide in the air, which converts the rain into dilute carbonic acid. Sulfuric acid is produced when sulfide rocks such as pyrite break open (because of either mechanical weathering or mining) and expose fresh surfaces to the air; the exposed sulfur becomes oxidized and dissolves in water to form dilute sulfuric acid. Another source of the acids that attack rocks is living material. A variety of corrosive organic acids are produced by microbes, and also by lichens growing on rock surfaces. Acids also come from the plants and invertebrate animals, both living and dead, that form the organic portion of soil.

The chemical weathering of rocks is greatly promoted if they have been mechanically weathered beforehand; the more fragmented the rock, the more surface there is for acids to work on. In time the products of both forms of weathering become jumbled together as a loose layer overlying solid bedrock; this is called *regolith*.

The most finely divided products of chemical weathering are, with one exception, divided much more finely than the products of mechanical weathering; the exception is rock flour, which is not produced in large quantities. The smallest, most abundant products of chemical weathering are clay particles, derived from feldspar, the commonest mineral on the earth's surface. The next most abundant, and notably coarser, are quartz crystals—that is, sand grains. Quartz is the most resistant to chemical attack of the common minerals, so it accumulates as a residue when any of the many kinds of rocks that contain it are chemically weathered.[7]

Most of the products of mechanical weathering are so much coarser than those of chemical weathering that it is not surprising that their subsequent fates are different too. For the most part, the products of mechanical weathering are shifted by mass wasting. Some of the products of chemical weathering are removed by erosion, and some dissolve in water and flow away with it. These processes combined carry weathered rock to lower elevations.

*Mass Wasting*

Of all the ways mass wasting happens, big landslides are the most spectacular. Whenever a mass of unattached or weakly attached rock chances to accumulate at the top of a steep slope, a landslide impends. The accumulated material, held precariously in place by friction, may be the product of long-continued weathering, or the debris of earlier landslides, or the ejecta of a nearby volcano. The poised mass awaits the conversion of its PE to KE. When something happens to trigger it, the mass starts to slide or fall, and the conversion begins.

The trigger may be a heavy rainstorm. For example, the excessively heavy rain accompanying Hurricane Mitch in October 1998 filled a lake in the crater of a dormant Nicaraguan volcano to overflowing; the escaping water, mixed with volcanic ash, created a mudslide that buried about 2,000 people. Numerous other mudslides caused by Mitch, together with flooding, brought the death toll to more than 11,000.

A landslide believed to be one of the earth's largest in several thousand years was the Frank Slide, which fell in April 1903 near the eastern entrance to the Crowsnest Pass through the Rocky Mountains, in southwestern Alberta.[8] Almost half the top of Turtle Mountain collapsed into the valley below, burying much of the little coal mining town of Frank; about 70 people were killed. The slide is thought to have been triggered by frost cracking on the grand scale, caused when large volumes of meltwater from a heavy snowpack poured into fissures on the mountain's summit; the rocks may have been weakened beforehand by coal mining at the foot of the mountain.

The weight of the Frank Slide has been estimated at $9 \times 10^{10}$ kg, and the distance it fell was approximately 1,000 m. With these numbers we can quickly calculate the amount of potential energy that its fall liberated (see chapter 2). Recall that the number of joules of energy is the mass (in kilograms) times the height of the drop (in meters) times the acceleration due to gravity, which is 9.81 meters per second per second, or 9.81 m s$^{-2}$. The energy of the slide was therefore

$$(9 \times 10^{10}) \times 10^3 \times 9.81 \text{ J, or approximately } 9 \times 10^{14} \text{ J.}$$

This enormous amount of energy was dissipated in about one hundred seconds. What happened to it?

Every time one rock strikes another, both are deflected and diverge in new directions and with altered speeds. Their combined speed, and therefore their combined KE, is reduced because both rocks are dented, broken, or chipped—which uses up some energy. If the collision is gentle these "deformations" may be too slight to be noticeable, whereas a more violent collision causes one or both rocks to shatter. In any case, chemical bonds are altered and heat energy is released, as is obvious when sparks fly.

By the time all the rocks come to a halt at the bottom of the fall, their gravitational PE has been lost because their elevations have been lowered; it could be restored if all the rocks were carried up to their original positions again—entailing much work! At the same time, more energy, including noise energy, has been generated by all the collisions and has been immediately dissipated as waste heat (entropy).

In some landslides the debris comes to an abrupt stop on reaching level ground, piling up into a big mound; in others the debris continues to move forward and doesn't come to rest until it has spread out over a large expanse of lowland. Such slides are known as *long-runout* slides. When a long-runout slide reaches the bottom of a valley, part of it may even keep on going, climbing some way up the opposite slope.

The Frank Slide is a famous example of a long-runout slide. After a rapid descent of 1,000 m, it "flowed" onward, across the Crowsnest River valley and 130 m up the valley's far side, leaving a sheet of shattered rock 30 m thick spread over the land. The leading edge of the sheet is 4 km from the base of Turtle Mountain; when you stand there, in a "sea" of broken limestone, it seems inconceivable that the debris came all the way from Turtle Mountain, 4 km away in the distance.[9]

Another well-known long-runout slide is the Elm Slide of 1881. It destroyed much of the Alpine village of Elm in Switzerland, killing 115 people. About $8.2 \times 10^{10}$ kg of rock slid 600 m down a mountainside, from which it follows that the potential energy lost in the slide was close to $5 \times 10^{14}$ J, a little more than half that of the Frank Slide. After reaching the foot of the mountain, the slide flowed on for 2 km before stopping; the average slope of the descent from start to finish was only 17°, much less than the normal angle of rest of piled rocks. As with the Frank Slide, some of the debris flowed uphill.[10]

The noteworthy character of long-runout slides is that their debris appears to flow like a liquid instead of behaving as you might expect a heavy mass of solid material to behave when it lands abruptly. For years the surprising "flow" of landslide debris was thought to happen when a cushion of air became trapped beneath the falling debris. But the debris of long-runout slides is observable on the moon, where the walls of some large lunar craters formed by meteorite impact have collapsed; trapped cushions of air cannot be the cause on the airless moon. The "fluidization" of dry rock debris is now believed to have the following explanation: In certain conditions, the mass of separate fragments forming the debris act like "molecules in a gas . . . . The entire collection of rocks [behaves] like a dense gas and . . . naturally, [flows] like a fluid."[11] The phenomenon is called *acoustic fluidization*.

When slide debris flows uphill at the end of its runout, as happened at Frank and Elm, it is imitating to a small degree the behavior of a glass marble released just inside the rim of a smooth bowl, which rolls to the bottom of the bowl and then nearly to the top on the opposite side. In both cases the potential energy of a mass (slide debris or marble) is converted to kinetic energy as it descends to the foot of a slope (a mountainside or the side of a bowl), after which momentum carries it upslope again, promptly restoring a portion of the PE it has lost. The rest of the energy is dissipated as heat and noise. The only difference between the two cases is that the proportion of PE conserved is far greater, and that of PE dissipated far less, for the marble in a bowl than for the rocks sliding into a valley.

Big landslides are soon over: huge quantities of debris complete their journey in seconds or minutes. Mass wasting in slow motion takes place too when individual rocks fall from a precipice one at a time and accumulate at the bottom; the commonest cause is frost cracking. The angular rocks pile up where they land, forming steep slopes of *scree* (also known as *talus* or *colluvium*) lying at the material's angle of rest, which is typically between 30° and 35°.

Scree slopes are common in mountainous country. The potential energy released while a scree slope builds depends on the mass of rocks in the scree and on the distance they have fallen. It is calculated in the same way as the PE of a sudden landslide; the amount of potential energy is the same whatever the speed of its release.

The slowest, least conspicuous form of mass wasting is *creep*. It happens on all slopes, however gentle. Obviously, any crumb of soil or particle of rock that chances to be dislodged on sloping ground will automatically move downs-lope, because the pull of gravity prevents it from moving upslope. Creep is the result. Crumbs of soil are dislodged in innumerable ways; they are scattered from the roots of wind-thrown trees; they are shifted by the movements of burrowing animals and pushed by the shoots of growing plants; they are raised and then let down a millimeter or two farther downslope every time the ground beneath them freezes and thaws. Wetting and drying expand and con-tract the soil just as freezing and thawing do, with the same result. Wind and rain displace both soil crumbs and rock particles.

A particle nudged out of place by moving air or water, however, is the ob-ject of erosion as much as of mass wasting. On a small scale, the two processes lose their distinctness.

## River Erosion

Erosion is the transport of weathered rock by moving fluids, either water or air. At the present time (geologically speaking) erosion by rivers is the princi-pal form of erosion.[12] The questions to be looked into are, How do rivers trans-port the material produced by weathering? What becomes of the material? And what are the energy exchanges?

The material to be transported consists of clay particles, sand grains, and cobbles, plus a small amount of rock flour that behaves like clay particles. These materials roll into rivers after "creeping" down adjacent valley slopes, or they are washed in by rain, or they come from clods of soil that fall into the water from a river's banks after a rainstorm and disintegrate. These last are re-cycled particles, as we shall see below.

Once they are in a river, the particles become *sediment*. The two chief com-ponents of the sediment, clay and sand, behave differently because of the markedly different sizes of their particles.[13] The fine particles, clay and silt (plus rock flour), are light enough to remain suspended in the water, whose turbulence supports them; they make the water muddy, and they are borne along with the water, whose speed is only slightly reduced.

The coarse particles, mainly sand grains, sink to the bottom because of their weight. If the current is gentle they become one with the riverbed, which more often than not consists of the same material. When the current speeds up, the sand at the bottom is swept along by the moving water as *bedload*.

A word on the subsequent fate of these sediments before we return to con-sider the energy of all these activities. River sediments are eventually carried out to sea and settle in layers on the seabed. This may not be the end of their travels: mass wasting goes on underwater as well as above it. Accumulating sediments often build up at the top of a slope on the seabed until they become unstable and collapse. If the particles are already cemented together, they collapse in big blocks in a process called slumping; if the sediments are still loose, they flow like a liquid down submarine slopes, eroding submarine canyons in doing so, and end up as layers on the ocean floor.

In any case, submarine sediments eventually come to rest, and as time passes (millions and tens of millions of years) loose sediments become solidified into sedimentary rocks: clay and silt become shale and siltstone, and sand becomes sandstone. Of the world's land area, 67 percent is covered by sedimentary rocks, of which shale and sandstone are the most abundant.[14] Sooner or later the submerged layers (strata) of sedimentary rock are raised above sea level again, as the internal energy of the earth deforms its crust (see chapter 15). The shales and sandstones become available for weathering again, and the resulting clay particles and sand grains (perhaps with some newly formed rock flour added to the mix) find their way into rivers. Another stage in the unending recycling of earth materials is under way.

*The Energy in Rivers*

We must now consider how a flowing river acquires, and dissipates, energy. Obviously water at high elevations—newly fallen rain, melted snow, and preexisting lakes, for example—has potential energy by virtue of its elevated position, just as rocks on elevated ground have. The energy is converted to kinetic energy when the water flows, or the rocks fall (or roll, or slide) downhill. The potential energy at any point on a river's course has two components, however: elevation energy and pressure energy. The elevation energy exists because of the elevation of the riverbed above sea level; it corresponds with the gravitational PE of a rock on a mountain slope, which could just as well be called "elevation energy."

The pressure energy depends on the water's depth; hydraulic engineers measure it as "head." To understand the distinction between elevation energy and pressure energy, visualize the following situation: suppose the water level in a river were to sink almost to zero at any point on its course; then the pressure energy at that point would be almost zero, and the river's flow would almost stop, regardless of the elevation of the dry riverbed above sea level, which would remain unaltered. This makes it intuitively obvious that the

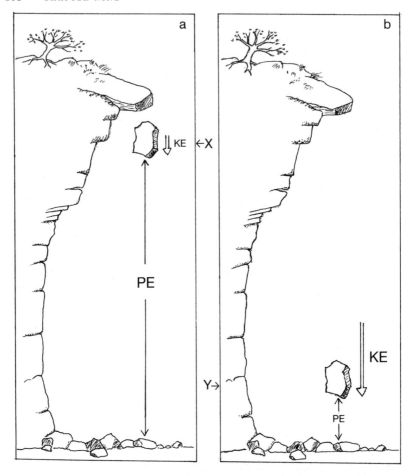

Figure 9.2. A rock falling from a precipice.(a) Earlier view with the rock at position X. (b) Later view, with the rock at position Y. Were it not for drag, the sum PE + KE would be the same at X as at Y. The open arrows represent the rock's velocity at each position, on which its KE depends.

depth of the water contributes to its total energy. For any site along a river, we may write the equation

$$PE = \text{elevation energy} + \text{pressure energy}.$$

Only a liquid can have pressure energy, not a solid.

Now compare the energy budget of a rock falling from a precipice with that

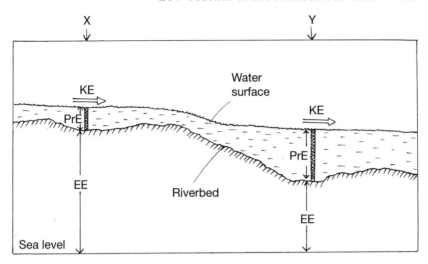

Figure 9.3. A river flowing in its channel (longitudinal section). The crosshatched strips represent "slices" of water at sites X and Y; were it not for drag, the sum EE + PrE + KE would be the same at both sites (see text for details).

of a "slice" of river flowing down its channel (see figs. 9.2 and 9.3). To avoid vagueness, we must choose arbitrary starting and stopping points to define the interval under consideration; these points are marked X and Y in both figures. We must also stipulate that the rock does not fracture and that the river in figure 9.3 neither gains nor loses water between X and Y.

At any instant, wherever it may be, each "object" (solid rock or liquid "slice") has a PE that depends on where it is at that instant and a KE that depends on its velocity at that instant. For both objects, PE and KE are changing continuously, from one instant to the next.

Were it not for friction (in the general sense, including drag), it would be true to say, for both the rock and the water slice,

$$\text{PE at point X} + \text{KE at point X} = \text{PE at point Y} + \text{KE at point Y.}$$

This follows from the law of the conservation of energy (see chapter 3).

In the real world, where drag is inescapable, the equation becomes

$$\text{PE at point X} + \text{KE at point X} = \text{PE at point Y} + \text{KE at point Y} + w,$$

where $w$ represents energy gone to waste between X and Y because of drag.

In their most concise form, these equations apply equally to the rock and the water slice. Now we put more detail into the equation for flowing water, taking note of the fact that the PE is the sum of elevation energy (EE) and pressure energy (PrE). Then in the ideal case, with no drag, the equation becomes (the word "point" is omitted for brevity)

EE at X + PrE at X + KE at X = EE at Y + PrE at Y + KE at Y.
(Do not confuse PrE with PE.)

This is the law of the conservation of energy as it applies to flowing water.

We can modify the equation for the realistic case (drag operating) in almost the same way, like this:

EE at X + PrE at X + KE at X = EE at Y + PrE at Y + KE at Y + $w_1$ + $w_2$.

Here $w$, the symbol representing waste energy—nonconserved energy—has been split into two parts, $w_1$ and $w_2$. The first, $w_1$, denotes *work* done by the flowing water, namely, picking up and transporting a load of sediment; the second, $w_2$, denotes true waste energy, namely heat and noise;[15] noise is thought to account for only one part per million of the total energy. Most of the energy goes into shifting bedload, which consists of all particles larger than very fine sand grains, of diameter 0.06 mm.

The lighter particles in bedload are dragged along by the flowing water. They move more slowly than the water, reducing its speed of flow, but they do not come to rest: turbulent eddies support them and keep them moving. The heavier particles, on the other hand, stop and start repeatedly as the strengths of the eddy currents fluctuate. Any particle of bedload that settles on the bottom is soon temporarily entrained (picked up) again: it may be dislodged by a minor eddy, or lifted by the flow of current over it in the same way that an aircraft wing is lifted, and raised into the faster current above the bed. After traveling a short distance, it will be deposited again. In this way the heavier particles hop forward without ever rising appreciably above the riverbed. Because it takes energy to lift the particles, a fraction of the river's kinetic energy is consumed.

The total energy used by all the rivers in the world in transporting sediments can be estimated approximately if we rely on another estimate, according to which erosion, chiefly by rivers, lowers the earth's land surfaces by 8 mm per century.[16] The earth's land area is about 149 million square kilometers, and the bulk density of the loose surface material left by weathering is about 1,500 kilograms per cubic meter. The mass of material removed from the surface is therefore about $1.8 \times 10^{13}$ kg. The average height of the land is about 875 m above sea level. With these data we can compute the potential energy lost, per century, in the same way as we computed the energy of the Frank Slide. The answer is

$$1.8 \times 10^{13} \times 875 \times 9.81 = 1.55 \times 10^{17} \text{ J per century.}$$

This is energy per unit time, in other words, power. It can be quickly converted to familiar watts (joules per second), or better yet, to watts per hectare, to make it easily imaginable. The answer is about one-third of a watt per hectare. Imagine the energy as light: it would amount to an invisible glow (if that *is* imaginable) on a dark landscape. The effects of the "glow" are cumulative, however. Over a few million years (not long, in geological terms), the "glow" creates and maintains most of the world's spectacular scenery. The folding and uplifting of the earth's crust by the earth's internal heat merely provides the raw material on which erosion acts.

The rate at which the earth's land surface is being denuded by erosion is vastly greater nowadays than it was before human undertakings like logging, farming, mining, and construction came to be practiced on an industrial scale. It has been estimated that the natural sediment load carried annually by the world's rivers is $1.6 \times 10^{13}$ kg, while the *un*natural load is $1.72 \times 10^{14}$ kg, more than ten times as much.[17] The long-term consequences of accelerated erosion could be as serious as other outcomes of the industrial-technological explosion, such as the increasing concentrations of greenhouse gases in the atmosphere. Time will tell.

Sediment transport accounts for only part of the energy dissipated by a flowing river, the part denoted by $w_1$ in the preceding equation. We have still to consider the part denoted by $w_2$, which is "wasted" or, equivalently, used up (converted to entropy) in overcoming resistance to the river's flow by the bed, the banks, and the sediment load itself. In a word, the flowing water experiences drag and is greatly slowed as a result.

Because of drag, no river can flow faster than its terminal velocity, the velocity at which the force of gravity accelerating it down its channel is exactly balanced by the force of drag restraining it (recall the account in chapter 5 of the terminal velocity of falling rain). The terminal velocity of any particular river depends on the straightness and smoothness of the channel walls. The importance of these factors becomes obvious if you mentally compare the gentle flow of a lowland river with the fierce jet of water spewing out of a penstock (sluice) bringing water from a dam to spin the turbines in a hydroelectric generating station.

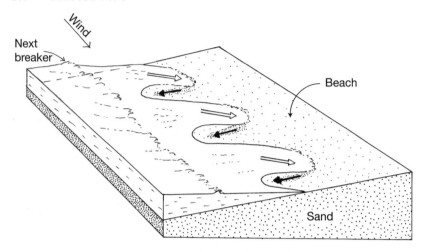

Figure 9.4. Beach drift. The open arrows show waves breaking obliquely and carrying sand obliquely upslope; the solid arrows show the direction of the backwash, carrying sand directly downslope. The resultant drift is to the left as seen from the beach.

The terminal velocity of a particular river depends also, of course, on the volume of water it happens to be carrying, which varies from time to time. A typical terminal velocity for a river of average size is in the neighborhood of 5 km/h, while the same river in spate could have a terminal velocity of 6 to 8 km/h. It is difficult to imagine the speed a river would have were it not for drag, but just thinking about it brings an appreciation of the inexorable increase of entropy going on wherever water flows.

## At the Beach Again

A river's current peters out when it reaches the sea. Whereas its suspended sediment drifts far from shore before settling to the bottom, the heavy bedload of sand is dumped much sooner and becomes the raw material for neighboring sand beaches. Its transport along the shoreline requires a further supply of energy.

Some of the needed energy comes from longshore currents, either tide currents or currents driven by the wind. Whatever their origin, currents in shallow water can transport a bedload of sand in the same way a river can.

Sand is transported along the beach itself by *beach drift*. It happens whenever wind-driven waves roll up a sandy beach obliquely, carrying some sand up with them; when they drain back by the steepest route down the beach, they carry the sand back into the water again a short distance downwind of the point where they picked it up (see fig. 9.4).

Quantities of sand are transported considerable distances by this inefficient process. Each successive wave gives a modicum of PE to its load of sand by carrying it up the beach, but the sand promptly loses the PE by rolling back down again. If the wind direction changes, the sand is carried back the way it came. Eventually the bulk of it is carried along the beach in the direction of the prevailing wind, but the energy lost to friction is enormous. This is obvious if quantities of fine pebbles are mixed with the sand; the roar of pebbles rolling and sliding over each other as they go first up and then down the beach slope is deafening, and noise represents only a minute fraction of the total energy lost, most of it as imperceptible heat. Not surprisingly, the quantity of energy lost is difficult to measure.

Ocean waves cause considerable mass wasting, too. Sea cliffs, formed by wave action, are subject to frequent slumps and landslides, triggered by the action of waves that undercut the cliffs. In this way coastlines are shaped by the energy of the sea, and every landslide liberates potential energy.

# 10 CHEMICAL ENERGY

*Searching for Chemical Compounds*

You have only to look around you, wherever you may be, to see
chemical compounds by the thousands, everywhere; all are substances consisting of atoms of two or more elements held together by chemical bonds. Nearly every solid material on
earth—nearly everything you see, animal, vegetable, or mineral—consists of chemical compounds, sometimes only one,
sometimes several together. Solid substances composed of single
elements uncombined with others are rarities; even "pure" gold
and "pure" iron are almost never absolutely chemically pure.

When elements combine to form compounds, energy is either produced (or liberated—"evolved," as it is often called) or
used up ("consumed"). A reaction that liberates energy, most
often in the form of heat, is *exothermic;* a reaction that will take
place only if energy is provided from some outside source is *endothermic.* Body warmth is produced by exothermic chemical
reactions: without them, rigor mortis would soon set in. The
earth's green plants, which directly or indirectly nourish almost

all other living things, grow as a result of endothermic reactions that use sunlight as their energy source. Chemical reactions affect every moment of our lives: one could say that life *is* a series of chemical reactions, forever consuming and liberating energy.

Chemical reactions are also going on, all the time and everywhere, in the nonliving world around us. If you leave a steel chisel outside all winter, you know it will be rusty by spring. The iron in the steel has combined with atmospheric oxygen to form iron oxide (rust). The reaction is exothermic: about 5,100 joules of heat energy are liberated for every gram of rust formed,[1] but the reaction happens too slowly for the heat to be noticeable except in carefully controlled laboratory experiments.

Chemical weathering of rock goes on all the time, as we saw in chapter 9. Minerals such as pyroxene, olivine, and hornblende, important ingredients of basalt, all contain iron that becomes oxidized—converted to rust—when fractured basalt is exposed to the air; rust is the chemical that makes iron ore red.

In the ordinary outdoor world, the commonest inorganic reactions in progress—those causing chemical weathering—attract much less attention than organic reactions taking place in living things. This is because weathering transforms a newly exposed rock surface so that it comes into equilibrium with its new environment, in contact with the atmosphere. The energy turnover in living material is much more obvious to human observers than are the very much slower inorganic reactions going on all the time around us. Both kinds of reactions consume and liberate energy: the contrast between them is in their speed.

## The Energy in Chemical Bonds

Chemical reactions consist of the creation of new chemical compounds from existing ones. To create new compounds from old entails making and breaking chemical bonds; this is the point where energy enters the picture: energy is absorbed or released whenever chemical bonds change. Therefore, before continuing, we must consider what chemical bonds are and how they work.

Chemical bonds hold together the atoms in a molecule and, likewise, stick molecules to each other. A *force* must exist to create a bond, as the word "bond" implies. In chemical bonds, the force is electrical. It is a force of nature in the same way gravitation is a force of nature: both are fundamental characteristics of the physical world that cannot, at present, be explained in terms of anything more fundamental.

This does not mean, however, that chemical bonds are all alike. They differ from each other in magnitude, depending on which elements the bonded atoms belong to, and they differ in kind—bonds operate in several ways.

The simplest kind of bond, though not the commonest, is an *ionic* bond. The bonds holding together atoms of sodium and chlorine to make sodium chloride (table salt), for example, are of this kind: in a laboratory experiment, suppose a sodium atom that has become a positive *ion* by losing an electron comes close to a chlorine atom that has become a negative ion by picking up a stray electron; because of the electrical attraction between the two ions, one positively and the other negatively charged, they unite to form sodium chloride. Each step entails an energy change. The net result is the liberation of energy: for every gram of sodium chloride formed, 7,000 J of energy, in the form of heat, are produced.

The chemical bonds that hold together the atoms in organic molecules are known as *covalent* bonds. A covalent bond exists whenever two atoms share a pair of electrons. Covalent bonds are not confined to organic molecules, and the atoms bound together need not be of different elements. For example, in three common gases, hydrogen, oxygen, and nitrogen, each molecule consists of a pair of atoms of the element concerned, united by a covalent bond; that is why the gases are written, in chemical symbols, as $H_2$, $O_2$, and $N_2$.

The simplest of all molecules, the hydrogen molecule, neatly illustrates the structure of a covalent bond. Each of its two atoms consists of one positively charged proton, the nucleus, and one negatively charged electron. The way they are covalently bound is diagrammed in figure 10.1. Figure 10.1a shows an isolated hydrogen atom; its single electron is somewhere in the circular cloud of dots surrounding the nucleus; one cannot say precisely where, just somewhere, moving rapidly, most frequently where the dots are densest. Figure 10.1b shows two hydrogen atoms united to form a molecule. The atoms share their two electrons: both are to be found somewhere in the oval cloud of dots encasing the two nuclei.

Free hydrogen in its natural state exists as molecules; to break all the bonds in a gram of hydrogen gas would require 217,000 J of energy. Breaking the bond in a single molecule uses $7.23 \times 10^{-19}$ J. This is the *covalent bond energy* for one hydrogen molecule. Obviously joules are inconveniently big units for use with individual molecules; the appropriate unit for these tiny amounts is the *electronvolt*, abbreviated to *eV* (for more on electronvolts, see chapter 16). One eV is the same as $1.6 \times 10^{-19}$ J; thus the bond energy for hydrogen is 4.5 eV. It is the energy required to break the bond in a hydrogen

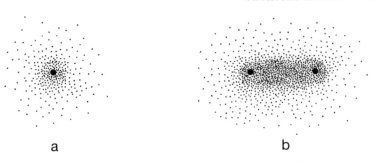

a                                        b

Figure 10.1. (a) A solitary atom of hydrogen. (b) A molecule of hydrogen. The denser the stippling, the greater the chance of finding an electron there at any given time. The big black dots are the nuclei.

molecule; equally, it is the energy liberated when two hydrogen atoms join to make a molecule.

Ionic bonds and covalent bonds normally hold together the atoms in a molecule.[2] Two other kinds of bonds attach molecules to each other; they are the bonds mentioned in chapter 9, which make solid objects stay solid, so that energy is needed to break them. They are not nearly as strong as the bonds joining the atoms in a molecule.

The intermolecular bonds that keep organic materials, including living organisms, from falling apart are *hydrogen bonds;* they form between a hydrogen atom in one molecule and an oxygen (or nitrogen or fluorine) atom in another. These three elements have atoms whose electrical characteristics enable them to make exceptionally strong bonds with hydrogen atoms. The commonest hydrogen bonds in living things are those in which oxygen is the element bonded to hydrogen.

Hydrogen bonds are the strongest intermolecular bonds, but they are not nearly as strong as covalent bonds, typically less than one-tenth as strong.[3] Weaker bonds linking molecules to each other exist as well, known as *van der Waals forces* and caused by local concentrations of electric charge on molecular surfaces. They are much less common in natural materials than in synthetics such as plastics.[4]

*Energy in Chemical Compounds*

The energy liberated or consumed in any chemical reaction is the  net result of the energies liberated and consumed when chemical bonds are made and broken. Complicated reactions like photosynthesis, for example, entail numerous makes and breaks to produce the end product, glucose, from its raw materials, water and carbon dioxide. The net result is consumption of energy; that is, the reaction is endothermic.

The energy consumed in photosynthesis is solar radiation, and the amount consumed is 16,000 J per gram of glucose created. The energy becomes stored in the glucose as chemical potential energy awaiting ultimate liberation as heat, in the same way that a rock poised at the top of a precipice stores gravitational potential energy awaiting liberation as movement. If you burn a gram of glucose, the photosynthetic reaction is reversed: water and carbon dioxide are produced and 16,000 J of energy are liberated—the potential energy is potential no longer.[5] Or instead of burning it, you could eat it: the chemical reactions that constitute digestion would then yield 16,000 J of energy for the acts of living, such as moving, growing, and keeping warm.

We have already mentioned the energy liberated in two simple exothermic reactions; recall that just over 5,000 J are liberated for every gram of rust produced by the oxidation of iron; and 7,000 J are liberated when sodium and chlorine combine to form a gram of table salt. It would be necessary to supply the same amounts of energy from an outside source to undo these reactions: 5,000 J to unmake a gram of rust and 7,000 J to unmake a gram of table salt.

The energy change that accompanies every chemical reaction is called an *enthalpy change*. At first this term seems redundant: If you mean an energy change, why not say so? The reason is that an exothermic reaction produces two kinds of energy. The first is *free energy*, or energy capable of doing useful work—heating water, for instance. The second kind is entropy. Recall, from chapter 3, that it is *never* possible to make all the energy supplied to a system do useful work; some is always dissipated in a useless form, as entropy. We noted above that it takes 16,000 J of solar energy to energize the photosynthesis of one gram of glucose, and the same number of joules are liberated if the glucose is burned. But the liberated joules cannot all be useful energy, or we should have the makings of a perpetual motion machine. That is why the 16,000 J are given the name *enthalpy*, which is made up of both free energy (the active or useful kind) *and* entropy.[6]

Up to this point we have described entropy as "waste heat." Alternatively, it can be described as the energy of the random motion—the "milling about"— of the molecules in every substance whose temperature is greater than absolute zero (0 kelvins). The energy of this milling about depends on the chemical nature of the substance as well as on its temperature. On average, the entropy of gases exceeds that of liquids, and the entropy of liquids exceeds that of solids. But this is true only on average. At room temperature, the entropy of solid table salt slightly exceeds that of liquid water; likewise the entropy of liquid alcohol exceeds that of helium gas. Solid substances have a wide range of entropies. For example, the entropy of lead is more than thirty times that of a diamond. This is because the atoms in a diamond are held firmly in their places in the diamond's crystal structure, whereas the atoms in a lump of lead are comparatively free to move; in brief, diamonds are well-knit, lead is rickety.

The change in enthalpy taking place in a chemical reaction is scarcely influenced by the temperature at which the reaction happens, but the partitioning of the enthalpy between free energy and entropy is strongly influenced: the ratio of free energy to entropy is much greater at low temperatures than at high ones.

## Ice and Steam

Enthalpy changes are familiar to everybody. When water freezes, or vaporizes, the change in the water is a change in enthalpy even though no chemical reaction, in the usual sense, has taken place. Likewise when ice thaws or water vapor condenses.

Suppose you hold an ice cube in your hand: heat passes from your hand to the ice—that is why your hand is chilled—but the ice is not warmed above freezing point. On the contrary, it stays at freezing point while it gradually turns to water. The ice gains enthalpy, but no part of the enthalpy gained in this case is free energy; it is all entropy, manifested in the greatly increased mobility of the water molecules swirling around as a liquid instead of being held rigidly in place in crystalline ice.

Similarly, suppose you leave a pan of water boiling on a hot plate. Because the water is already boiling, its temperature will not rise any higher, but at the same time energy is passing from the hot plate to the water, changing the water's enthalpy. The change, an increase, has no free energy component; it consists wholly of an increase in entropy, manifested in the much greater mobility of the water molecules when they become a gas—the vapor rising from the boiling water—than they had as a liquid. If the water is boiled in a kettle with a loose lid, some of the enthalpy is free energy: it rattles the lid.

The change in enthalpy when a gram of ice melts is 335 J, and the change when a gram of water vaporizes is 2,259 J. Converting the joules to calories—the more familiar units in this context—gives 80 cal and 540 cal, respectively. These numbers will be recognized by many as the *latent heat of freezing* and the *latent heat of vaporization* of water. Indeed, the foregoing discussion deals with the same topic as the section on water vapor and energy transfers in chapter 5. Here we have seen how freezing and melting, or vaporizing and condensing, behave exactly like chemical reactions so far as energy transfers are concerned.

## Chemical Energy to Electrical Energy and Back Again

To describe chemical reactions as endothermic or exothermic, according as they consume or liberate energy, is somewhat misleading: it suggests, falsely, that the energy involved when chemical bonds are made and broken always takes the form of heat. More inclusive terms are *endergonic* and *exergonic*. The former describes a reaction that must be supplied with free energy of some kind (not necessarily heat) from an external source if it is to proceed, the latter a reaction that proceeds without such a source. Note that it is the direction of flow of free energy, as opposed to enthalpy, that determines whether a reaction is endergonic or exergonic.[7]

Some chemical reactions entail the liberation or consumption of electrical energy. An obvious example is the charging and discharging of the lead storage battery in an automobile. This is the way it works.[8] The battery consists of plates of pure metallic lead alternating with plates of lead oxide; there are spaces between the plates. All the lead plates are wired to a single conducting cable ending at the battery's negative terminal; likewise, all the lead oxide plates are wired to a cable ending at the positive terminal. The sheaf of spaced plates is immersed in a mixture of sulfuric acid and water.

Every time the circuit is completed (for example, when you switch on the ignition or the headlights), a chemical reaction starts in the battery. The lead plates react with the sulfuric acid to produce lead sulfate and free electrons, and the electrons start to stream through the circuit, producing an electric current. The flowing electrons do the required task—turning the starter motor, say, or lighting the headlights—and then return, via the positive terminal, to the lead oxide plates; there they combine with the lead oxide and some of the sulfuric acid to form more lead sulfate. When the battery is in use, lead sulfate accumulates on both kinds of plates while the sulfuric acid is gradually used up. As a result, the acid-and-water mixture becomes more dilute and, consequently, less dense; its density is what is measured when the battery is tested with a hydrometer.

When the acid becomes too dilute for the reaction to continue, the battery is said to be discharged. It can then be recharged by passing an electric current from some other source through it; this drives the chemical reactions in reverse, restoring things to their original condition: that is, the electrical energy supplied is converted back to chemical energy, available for reconversion to electrical energy when it is required.

The salient feature of what happens in an electrical storage battery is the transference of electrons from one material, lead, to another, lead oxide. Vast numbers of chemical reactions are of this kind: in simple ones, electrons are released by a pure element of one kind and become attached to a pure element of another kind. All these reactions—both simple and complicated—are called *redox reactions.*[9] A redox reaction gives off energy when it goes in one direction (as when a battery is giving an electric current) and absorbs energy when it goes in the opposite direction (as when a battery is being recharged by an electric current fed into it). Redox reactions are what make living bodies live; biochemical redox reactions entail the transference of electrons between large, complicated organic molecules, but they are redox reactions nonetheless. They are what is happening when food is converted to energy; the electrical energy yielded by the reactions is transformed into the mechanical energy of movement and into heat.

All transfers of biochemical energy involve electron transfers, and they usually take place in several steps: the energy yielded by one reaction powers a second reaction that would not happen without it, which powers a third reaction, and so on. The chemical that functions most often as an intermediary in these sequences of reactions is adenosine triphosphate, well known by its acronym ATP; it has been called the principal carrier of biological energy.[10]

The electron transfers in biochemical reactions do not produce strong, easily detectable electric currents. Reactions having the same effects as the discharge and recharge of an automobile battery are much less common in nature than in the technological world. The only conspicuous examples in nature are the strong electric shocks delivered by certain species of fish when they are disturbed: the electric eel of South American rivers, the electric catfish of central Africa and the Nile, and several species of electric rays living in tropical and temperate seas in various parts of the world. All of them are capable of electrocuting a human being when their "batteries" are fully charged.

# 11 ENERGY ENTERS THE BIOSPHERE

*The Light That Gives Life*

All living things, without exception, must have energy if they are to survive, to grow, and to multiply. And all living things, with a few exceptions, obtain this energy, directly or indirectly, from sunlight. The few exceptions will be described in chapter 12. The guarded phrase "directly or indirectly" allows for the fact that only some life forms—green plants—are able to convert solar energy directly into chemical energy. Other organisms obtain their solar energy at second hand (or third or fourth, or . . . ) by eating the green plants that first captured it, or by eating the animals that ate the green plants, and so on back through the whole series of organisms forming a *food chain*. Food chains themselves are usually connected to other food chains by lateral and diagonal links, to make *food webs;* to keep the discussion clear of needless complications, in what follows we concentrate on single food chains.

Green plants, as everybody learns at school, are the factories that capture energy from sunlight in a series of chemical reactions collectively known as *photosynthesis;* the green pigment *chlorophyll* always takes part in the reactions. Before going, rather sketchily, into the chemical details, it is worth considering the energy that drives them—sunlight.

Sunlight—more formally, solar energy—consists of electromagnetic waves; the way *they* convey energy is the topic of chapter 18. The point to notice here is that electromagnetic waves are not all alike; on the contrary, they have a tremendous range of wavelengths. The shortest waves reaching the top of the atmosphere from the sun in appreciable amounts are about 0.1 micrometers long, the longest close to 4.0 micrometers. A micrometer (symbol $\mu$m; $\mu$ is the Greek letter *mu*) is one millionth of a meter.[1]

About 90 percent of the *solar spectrum* lies between these limits. As we shall see, much of the radiation reaching the top of the atmosphere never penetrates to ground level. Nearly all that does is visible to humans; it constitutes the *visible spectrum,* the wave band of electromagnetic wavelengths to which human eyes are sensitive; it ranges from 0.40 $\mu$m to 0.71 $\mu$m.

The visible spectrum for humans, seen in its entirety, appears as white light. For human observers, different wavelengths produce light of different colors; the shortest ones appear violet, the longest, red. All this is well illustrated by a rainbow, which splits white light into its component colors or wave bands (how it all looks to other species is outside the scope of this book).

What is not so obvious when you look at a rainbow is the pronounced difference in energy content of the wave bands that constitute the different colors. The solar spectrum between 0.1 $\mu$m and 2.0 $\mu$m is shown in figure 11.1. The total area of the spindle-shaped strip represents the total energy, between these limiting wavelengths, received at the outer limits of the earth's atmosphere. The "tails" of the spectrum—at wavelengths less than 0.1 $\mu$m and greater than 2.0 $\mu$m—have been cut off to keep the figure compact. Consider the width of the strip, disregarding the shading; the width at each level represents the energy at the relevant wavelength, as shown on the scale at the left. It is obvious that a large proportion of the total energy (about 41 or 42 percent) is in the visible spectrum. Most of the incoming radiation with wavelengths outside the visible spectrum is absorbed in the atmosphere and never reaches ground level. The wave bands absorbed, or partly absorbed, are shown black (the few narrow wave bands in which absorption is only partial are not distinguished in the figure). The short wave, ultraviolet radiation, with wavelengths less than 0.34 $\mu$m, is absorbed high in the atmosphere, by oxygen, $O_2$, and by the ozone, $O_3$, of the famous ozone layer, which saves us all from being seriously sunburned. The long wave, infrared radiation, with wavelengths

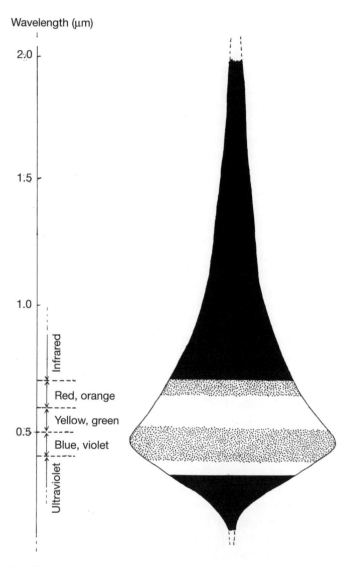

Figure 11.1. The solar spectrum at the top of the atmosphere (the energy tapers off gradually beyond the limits shown here). The width of the spindle-shaped figure at any level is proportional to the energy at that level; wavelengths in micrometers, μm, are shown on the vertical scale at the left. Wave bands absorbed or partly absorbed in the atmosphere are black. The photosynthetically active wave bands are stippled.

greater than 0.71 µm, is absorbed lower in the atmosphere, mostly within 10 km of the ground, by carbon dioxide and water vapor, the two chief "greenhouse" gases.

This leaves only radiation in the wave band 0.34 to 0.71 µm unabsorbed and able to reach the earth's surface; most of the energy in this region of the spectrum is visible: only that between 0.34 and 0.40 µm is ultraviolet, invisible to humans but visible to bees.

Now observe the two stippled segments in the unblackened part of the spectrum. They represent the wave bands absorbed by the chlorophyll in green plants or, equivalently, the energy used in photosynthesis.[2] Photosynthesis consists of a long, complicated series of chemical reactions, so it isn't surprising that the necessary radiation is not confined to a single narrow wave band. The two wave bands bearing the energy essential to nearly all life on earth are 0.40 to 0.50 µm (violet to blue) and 0.65 to 0.70 µm (orange to red).

The atmospheric absorption of sunlight we have so far considered goes on steadily, changing only gradually as the composition of the atmosphere changes; global climate change, whatever its cause, gives persuasive evidence to both scientists and nonscientists that the atmosphere *is* changing.

In addition, rapid fluctuations in the amount and kind of sunlight reaching any given spot of ground go on all the time. Think of the contrast between day and night, between winter and summer, between high latitudes and low. Think of the effect of clouds and haze. Natural aerosols—the water droplets in clouds and fog, and the myriad fine particles creating haze—dust, sea salt, volcanic ash, and bacterial spores—scatter sunlight as well as absorbing it. Changes in the brightness of the light, and in its spectral composition, affect the rate of photosynthesis. Surprisingly, the diffuse white light of a hazy day contains a higher proportion of photosynthetically active energy than does unobstructed sunlight, even though its total energy is less.[3]

The amount of usable energy reaching vegetation from the sun is difficult to measure directly because the rate at which it arrives is forever changing: it is most easily estimated by measuring the rate at which plants grow, known as their *productivity*.

## Ecological Productivity

The chemical reactions of plant photosynthesis can be summarized thus: carbon dioxide + water vapor + solar energy $\Rightarrow$ glucose + oxygen. This statement shows only the raw materials (to the left of the arrow) and the final products (to the right); it skips all the intermediate steps. Glucose is a simple carbohydrate; glucose molecules are the building blocks for a host of more elaborate carbohydrate molecules, among them cellulose, the most abundant organic chemical in nature and the one that, apart from water, constitutes the bulk of all plant material.

The individual reactions omitted from the statement are enormously complicated; fortunately they need not detain us. We are concerned with the capture of solar energy by the earth's vegetation, a process known as *primary production* because it is the stage at which the sun's energy is captured by living things for the very first time. The production of flesh by plant-eating animals—of beef by grazing cattle, for instance—is secondary production.

The energy required to produce one gram of glucose by photosynthesis is 15,650 J. A few words on units are necessary here. In the past (and to some extent still in the present), research workers studying ecological energetics, as it is called, used calories or kilocalories (1 kcal = 1,000 cal) as the unit for energy. As we saw in chapter 3, the relation between these units is 1 calorie = 4.186 joules, often rounded to 4.19 J. In traditional units, the energy required to produce one gram of glucose could therefore be given as 3.74 kcal.

Dietitians have confused matters by using the word Calorie (with a capital C) to mean kilocalorie. If you are told that your daily food intake ought to be about 2,000 Cal, the amount meant is 2,000 kilocalories, or approximately 8,400 kilojoules. If you ask someone who doesn't know the difference between Calories and calories, you might be told 2,000 calories (and you would starve). And if you ask by phone, the reply may be uninterpretable.

Because of this astonishingly ill conceived and carelessly used measuring system, it is much safer to follow the physicists and use joules (J) or kilojoules (kJ) as energy units.

The energy captured in one growing season by photosynthesizing plants is usually measured by harvesting the plants, drying them, and weighing the "dry matter." It is generally assumed that a gram of dry matter contains 0.45g of carbon, and that 10 kcal, or 42 kJ, of solar energy are captured per gram of carbon incorporated in the dry matter.[4] This is equivalent to saying that each gram of dry matter represents 4.5 kcal of energy, or 18.9 kJ. The conversion factors are only estimates and should not be expected to give exact results every time; but if the factors are used repeatedly, overestimates and underestimates tend to cancel out.

The result obtained by harvesting, drying, and weighing the plants on a chosen area and then converting this measured weight to the energy equivalent gives an estimate of what is known as the *net primary production* (or NP1) for the vegetation in the environment concerned. In the symbol, NP stands for *net production,* and the 1 indicates that it is *primary.*

It is net in the sense that it doesn't represent all the energy the plants have absorbed. Plants need additional energy, first, because they must photosynthesize for immediate nourishment as well as for growth and reproduction and, second, because they need some of the sun's heat energy in addition to the light energy used to drive photosynthesis. We consider these two extra energy inputs in turn.

First the "extra" photosynthetic energy: while plants are putting on weight by growing new tissue, they have to keep themselves alive, and to do this they must consume a fraction of the weight they have put on instead of storing it as new growth. This they do in the process of respiration, which "burns" carbohydrate "fuel" to produce the energy needed to maintain life. The total energy absorbed, both that used for new growth and reproduction and that used for maintenance, is *gross primary production* (GP1). The whole business of growing and maintaining life while doing so can be likened to running up a "down" escalator. Gross production is represented by the sum total of the heights of all the steps climbed in a given time; respiration is represented by the distance the escalator descends during this time; net production is represented by the actual height the climber gains—equivalently, by the difference between the gain and the loss.

Consider now the second kind of "additional" energy plants require: heat from the sun. Although photosynthesis creates all the organic matter in a plant, the plant also requires water and some essential inorganic minerals. The plant absorbs soil water, with the minerals dissolved in it, through its roots. To keep the flow moving, the water entering via the roots is finally "exhaled" through the leaves, after giving up its dissolved minerals on the way. The "exhalation" is called *transpiration:* water vapor evaporates from the interior cells of the leaves, first through their thin cell walls into spaces within the leaf and then to the outside air via *stomata,* tiny perforations in the comparatively tough outer "skin" of the leaf. Note that the process entails evaporation, which requires heat energy from the sun.

The sun's electromagnetic radiation thus provides two kinds of energy for plant growth: light energy for photosynthesis and heat energy to keep the plants' water circulation going. The heat is as important as the light.[5] The difference in the vegetation of different latitude zones is determined at least as

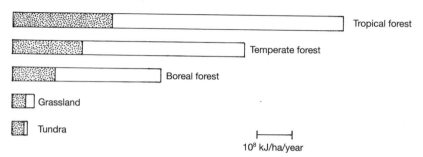

Figure 11.2. Typical productivities for five ecosystems. For each, the length of the whole bar represents the gross primary production, GP1; the stippled section represents the net primary production, NP1. Note the scale bar at the bottom representing $1.0 \times 10^8$ kilojoules per hectare per year (1 hectare = 100 × 100 m).

much by the sun's heat, which controls the water circulation rate, as by the sun's light, which powers photosynthesis.

Now back to the products of photosynthesis. The magnitude of the GP1, and the proportion of it stored as NP1, varies widely among ecosystems. Figure 11.2 shows some typical examples.[6] The bars show the magnitudes of both GP1 and NP1, per square kilometer, for five ecosystems. The proportions of the GP1 stored as NP1 are almost the same (0.3) in the three types of forest, in spite of the markedly different climates they grow in. The proportions stored by the low vegetation of middle and high latitudes are twice as great: 0.6 in grasslands and nearly 0.7 in tundra. Life is slow in the cold.

The total quantity of solar energy captured in a year and temporarily stored by all the world's terrestrial vegetation has been estimated at about 1.9 × 10¹⁸ kJ.[7] This is the terrestrial NP1 of the whole world. This energy is captured by 1,840 billion metric tons (dry weight) of plants, the world's entire "standing crop" of land vegetation.[8] Note that the NP1 is easier to estimate than the GP1. It is only necessary to cut, dry, and weigh a sample of vegetation at the end of the growing season to determine the NP1 of that particular sample, whereas to find the GP1 it would be necessary, in addition, to estimate the amount of weight loss owing to respiration during the same season, a much more difficult task.

The final statistic to be examined here is the *efficiency* of photosynthesis. The question is what proportion of the solar energy absorbed by green plants is converted, by photosynthesis, into chemical energy. The answer depends on whether you measure the captured energy (the GP1) as a proportion of the available *photosynthetically active* energy, in the violet-blue and the orange-red wave bands, or as a proportion of the available energy of all colors, regardless of usefulness, which is about twice as great. Measuring the energy in the GP1 as a proportion of the total energy received, photosynthetic efficiency is typically in the neighborhood of 1 percent and rarely greater than 3 percent.

*Primary Productivity of the Oceans*

The "vegetation" of the sea, as well as that of the land, takes part in converting solar energy to chemical energy by photosynthesis. As with land plants, photosynthesis requires light, water, and a source of carbon, but the carbon need not all come as carbon dioxide. Carbon dioxide does dissolve in water to some extent, but a source more useful to most marine plants is bicarbonate compounds dissolved in the water.

Two entirely different groups of plants grow in salt water; one group consists of plankton species, the other of seaweeds. *Plankton* is the collective name for the swarms of tiny organisms that live all or part of their lives floating in the ocean and drifting with the currents. The individual organisms of the plankton range from insubstantial jellyfish down to microscopic and submicroscopic one-celled organisms including bacteria. A portion of them, collectively known as *phytoplankton*, contain chlorophyll and carry on photosynthesis, thus feeding themselves; they also serve as fodder for the *zooplankton*, the nongreen members of the plankton, which cannot feed themselves.

In the open ocean, practically all green plants are phytoplankton; the only notable exception is sargasso weed, a seaweed that floats at the surface far from land. Seaweeds in general, forming the second component of the marine vegetation, are confined to very shallow waters and, though not rooted, live attached to shoreline rocks, anywhere from the high tide line down to as far below the low tide line as sufficient sunlight penetrates. Seaweeds feed themselves by photosynthesis; many species don't look green (for example, kelps and rockweeds are brown and so-called Irish moss and dulse are purplish red) because the green of their chlorophyll is masked by other pigments.

The phytoplankton of the open ocean, most particularly the abundant phytoplankton of the ocean's upwelling zones, is more than four times as productive as the seaweeds worldwide.[9] The total NP1 of all the green plants in the world's oceans (phytoplankton and seaweeds combined) is about $1.1 \times 10^{18}$ kJ per annum, or slightly more than half that of all the terrestrial vegetation, even though the oceans cover 70 percent of the earth's surface.[10]

This is astonishing at first when you compare the total biomass—the dry weights of the "standing crops"—of the land plants and the marine plants doing the producing. Estimates of these quantities are 1,840 billion metric tons for land plants (as already mentioned), and 4 billion metric tons for marine plants.[11]

Consider the land:ocean ratios of these quantities. They are

$$\text{land plant biomass:marine plant biomass} = 1{,}840{:}4 = 460{:}1$$

and

$$\text{land plant NP1:marine plant NP1} = 1.9{:}1.1.$$

Why the spectacular difference?

The reason is that land plants grow exceedingly slowly compared with plankton organisms, or *plankters* as they are conveniently called. Land plants have life spans ranging from one year (for annuals) to hundreds of years (for forest trees). Plankters have life spans of days or weeks at most. Therefore, when you look at an expanse of land plants in the fall, you see all the growth they have made in the growing season just finished in addition to what was already there when growth started in spring; in other words, the year's NP1 is all present before you—or nearly all. Some twigs may have broken, some leaves may have fallen, and some grass blades may have been grazed. But a sample of living plankters will contain only what has been produced in the past few days or weeks; the rest of the year's NP1 is missing—it is either already dead or not yet born.

This also explains why the biomass of terrestrial vegetation is so much greater—460 times greater—than the biomass of all living marine plants. On land, plant material accumulates and persists; at sea it is transitory and quickly disappears.

The volume of water inhabitable by phytoplankton is not so great as the huge volume of the oceans at first suggests. Sunlight penetrates only the topmost layer of the ocean, as we saw in chapter 6. The brightness of the light drops off at increasing depths below the surface, quickly at first and then at an ever decreasing rate.

Excessively strong sunlight inhibits photosynthesis. Therefore, going down from the surface on a sunny day, to begin with the photosynthetic rate *in-creases* as the light loses intensity. The rate reaches a maximum at the level where the energy of the sunlight has decreased to about half of what it is at the surface; the depth, in clear water, is between 2 and 3 m.[12] At progressively greater depths, with the light growing dimmer and dimmer, the photosynthetic rate decreases rapidly.[13] Below about 10 m, the light is entirely blue, all other colors having been absorbed (see chapter 6); the unabsorbed wavelengths are in the photosynthetically active wave band, however.

At the level where the energy in the sunlight has dwindled to about 1 percent of its full intensity, the *compensation level* is reached. This is the level where the rate at which energy is captured by photosynthesis balances the rate at which it is lost by respiration. It is the level where, in terms of the escalator analogy, the climber is running up at the same rate as the escalator is moving down. Consequently the NP1 is zero. The level may be deeper than 100 m in exceptionally clear water.

Now compare the volume of water occupied by photosynthesizing phytoplankton in very clear water with the volume of air occupied by evergreen coniferous forest on fertile soil, given equal areas of the two contrasted ecosystems. Rather surprisingly, the volumes occupied turn out to be about the same. This is because their vertical extents are similar: photosynthesis takes place only in the topmost 100 m of clear ocean water, and the average height of full-grown coniferous trees is usually in the neighborhood of 100 m.

Before leaving the topic of primary production, it's worth noting that not all photosynthesis follows the formula for *plant* photosynthesis given at the start of the preceding section. The word *plant* used there is not redundant; it is used to distinguish the process from *bacterial photosynthesis.*[14] This happens in deep, clear lakes, clear enough for some sunlight to reach the bottom. So-called sulfur bacteria must be present; their habitat is deep  fresh water. Oxygen must be wholly absent—it poisons them. And hydrogen sulfide, supplied by decaying material in the mud of the lake bed, must be present. Given these conditions, the bacteria create carbohydrate ($CH_2O$), producing sulfur (S) in place of oxygen ($O_2$) as a by-product; the sulfur forms granules inside the bacteria before being used in further reactions.

Except that it depends on a slightly different form of chlorophyll, the photosynthetic reaction appears similar to that in ordinary green plants, with sulfur taking the place of oxygen. The "crude" formulas for the two processes, showing only the initial inputs and final outputs, and concealing pronounced differences in the intermediate steps, are

$$CO_2 + H_2O + \text{light energy} \Rightarrow (CH_2O) + O_2 \text{ in ordinary plants}$$

and

$$CO_2 + 2H_2S + \text{light energy} \Rightarrow (CH_2O) + H_2O + 2S \text{ in sulfur bacteria.}$$

*Secondary Production: Energy Climbs the Food Chains*

Solar energy captured and converted into chemical energy by photosynthesis has merely begun its journey through the biosphere. It is in the "bottom link" of innumerable food chains, along which are transferred the total energy requirements of all the world's animals.

The several links along a food chain are known as *trophic* (feeding) *levels*, and all the species in an ecosystem belong to at least one trophic level. All the world's "vegetation"—green plants including seaweeds plus phytoplankton— belongs to the first trophic level. The second trophic level comprises all the herbivores, the third level all the carnivores that eat herbivores, and the fourth level (when there is one) all the carnivores that eat the carnivores that eat the herbivores. Carrion feeders, or scavengers, belong to the level they would occupy if they killed their prey for themselves.

Here are two representative examples of four-link chains: on the Atlantic coast, seaweeds, forming level 1, are eaten by sea urchins (level 2), which are eaten by lobsters (level 3), which are eaten by humans (level 4). On the subarctic tundra, grass and seeds (level 1) are eaten by ground squirrels (level 2), which are eaten by weasels (level 3), which are eaten by golden eagles (level 4).

All this is elementary, and far too simplified to be useful in working out the energy budget of an ecosystem. In the first place, organisms at a high trophic level often eat food from several lower levels: humans eat lobsters *and* sea urchins *and* seaweed; golden eagles eat weasels *and* ground squirrels. It follows that humans and golden eagles—indeed, most "top" carnivores—belong to two or three trophic levels: they are often called (with some exaggeration) omnivores. It is even possible for a plant to occupy two levels. Some species of plants (mistletoe is an example) have insufficient chlorophyll to photosynthesize all the carbohydrate they need, and they obtain the rest of it by parasitizing other plants. Totally parasitic plants—for instance, coralroot orchids— belong squarely in trophic level 2. They are herbivores.

Rarely, a food chain loops back on itself. For example, a pitcher plant performs photosynthesis and so belongs to level 1. It also ingests the insects it captures, making it a member of level 3 when it consumes plant-eating insects and of level 4 when it consumes bloodsuckers like mosquitoes.

In working out the energy flow through ecosystems, all these fascinating minutiae have to be disregarded. In devising an energy budget for an ecosystem, trophic levels are the units considered, not specific groups of plants and animals. A particular trophic level, level 2 for example, does not consist of a specifiable group of animal species that are confined to that level. Rather, it consists of all the animals (and plants) whose food (energy) comes from level 1 in a single step; that is, herbivores together with omnivores (such as bears) that eat both plants and herbivores. Likewise, animals in level 3 derive their energy from level 1 in two steps; and so on. Recall that level 1 is where solar energy is first converted to the chemical energy that provides the energy in food.

The biomass belonging to level 1 in an ecosystem is the mass of all the photosynthesizing plants in the ecosystem. The biomass belonging to level 2 is the sum of the biomasses of all the strict herbivores *plus* the proportion of the biomass in carnivores that they obtained by eating plants; and analogously for successively higher levels.

Once the data have been gathered, a flowchart patterned like that in figure 11.3 can be drawn, showing how energy is transferred in an ecosystem. Note that the flow is upward. The chart applies equally well to a terrestrial or an aquatic (marine or freshwater) ecosystem. Although in describing the chart we speak of material objects such as plants, animals, food, and leftovers, in every case these materials contain captured solar energy; whenever quantities of biomass are measured, they can always be converted to joules or kilojoules.

The bottom panel of the figure shows a year's events in trophic level 1. All the living organisms at this level (they are all plants) are represented by the grass in the drawing. The gross primary production yielded by the plants is shown by the heavily outlined box labeled GP1. Note that the plants producing the GP1, or most of them, are perennial and go on living to produce again in future years; the perennial parts are not part of the GP1. As we saw earlier, the GP1 has three components, shown by the small boxes enclosed in the large box: E1, the plant tissues eaten by animals at level 2 and higher levels; A1, the additions—new growth and offspring—gained by level 1; and R1, the loss due to respiration by level 1. Note that E1 and A1 combined amount to the net production of level 1, otherwise the net primary production NP1; also, that R1 + NP1= GP1.

Now consider the "crumpled" box, D1. It is the detritus (or "waste," or "litter") pertaining to level 1. It consists of old plant fragments such as fallen leaves, broken twigs and branches, and any other bits of plants that grew in previous years and became withered or detached this year; plus that fraction of the year's growth (A1)—leaves, twigs, bark, fruits, and seeds—left uneaten; plus what has been eaten but not assimilated by animals at higher trophic levels. In a word, it is level 1's leftovers and higher levels' feces, which contribute nothing to any level in this food chain.

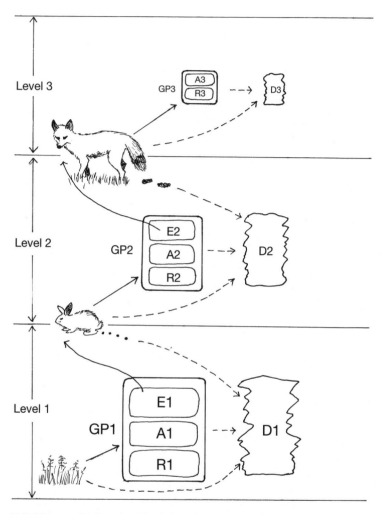

Figure 11.3. Diagram of a three-level food chain (read it from the bottom upward). The code letters are followed by the level's number. The standing crops at levels 1, 2, and 3 are symbolized by the grass, the rabbit, and the fox. Arrows lead from each symbol to an outer, partitioned box, GP (the gross production) and a "crumpled" box, D. The energy of GP is used in three ways (the small, inner boxes): E, eaten by animals at a higher trophic level; A, added, by growth and reproduction, to the biomass of the level that produced it; and R, lost through respiration. D is detritus ("litter" or "waste"), destined to decay or burn in the future. Note that $A + E = NP$, the net productivity of the level, and that $A + E + R = NP + R = GP$.

Leftovers don't accumulate, however. They are the basis for other food chains, called *detritus food chains*, to be considered further in chapter 12. It's impossible to emphasize too strongly the importance of detritus food chains in cycling energy and materials through the biosphere. In any ecosystem, the bulk of all production winds up as detritus. In this chapter we will do no more than list some of the ingredients in the detritus of different levels.

Back to level 2 of the flowchart, the herbivores: the energy of the standing crop of herbivores (represented here by the rabbit) comes from NP1; in other words, *all* the energy at level 2 is one step removed from solar energy. This energy goes to the destinations shown: E2 is eaten by carnivores; A2 is the herbivores' new growth, in the form of both increased size and offspring; R2 is their loss from respiration. The detritus at this level, D2, consists of dead, not yet decayed herbivore bodies, shed herbivore parts such as deer antlers, molted hair and feathers, and the shed outer skins of growing invertebrates (crustaceans such as crabs and lobsters, metamorphosing insects, and many more); also in D2 is undigested herbivore flesh and bones in the feces of carnivores.

Level 3 (represented by the fox) is the top level in this ecosystem, which accounts for the absence of E3. In other respects, level 3 matches levels 2 and 1. All the energy at level 3 is two steps removed from solar energy.

An interesting exercise for any naturalist is to visualize the flowchart for an aquatic ecosystem, in either fresh or salt water. The diagram would look the same except for the pictured organisms representing each level. The ingredients of the detritus would be very different. It would also be apparent that marine ecosystems tend to have a larger number of trophic levels than terrestrial ones and (if quantities were being measured) that the amount of energy dissipated by respiration is greater in warm-blooded than in cold-blooded animals because the former have to generate heat to stabilize their temperatures.

Another contrast between marine and terrestrial ecosystems that would emerge is a striking difference between the relative proportions of "producers" (level 1) and "consumers" (all levels above 1) in the two kinds of ecosystems.[15] In terrestrial ecosystems, the biomass of the consumers is only about one thousandth of the biomass of the producers—land vegetation. In marine ecosystems, the biomass of the consumers is about *twenty times as great* as the biomass of the producers—seaweed and phytoplankton.

The striking difference between land and sea in their ratios of producers to consumers is easy to visualize. About one-third of the world's land surface is forested, and another third is grassland; the biomass of all this vegetation is, understandably, orders of magnitude greater than the biomass of all terrestrial animals. The seeming abundance of consumers relative to producers in the ocean is equally obvious to the mind's eye. Except in a narrow zone near shore, the producers consist of swarms of tiny phytoplankters floating in the illuminated surface waters and seldom affecting its transparency to any noticeable degree; most of the consumers are carnivorous fishes belonging to several trophic levels, with the individual members of each level generally outweighing those at lower trophic levels, which they prey on. The biggest animals are whales.

The reason for the difference between land and sea in the ratio of producers to consumers was explained above in a different context: recall that plant material on land, especially trees, persists for decades or centuries, growing bigger all the time; compared with trees, land animals—even bears and moose—are comparatively tiny and have life spans that rarely exceed twenty years. At sea this order is reversed: plankters have life spans of days or weeks, whereas fishes at the top of their food chains, and also whales, live and keep on growing for many years.

Last, we come to the efficiency of energy transfer from each trophic level to the one above it in an ecosystem. The subject has been studied in tremendous detail, and not surprisingly the studies have supplied a torrent of numerical measures of efficiency for numerous ecosystems. The efficiency of the transfer from level 1 to level 2, say, is defined as the energy in GP2 measured as a fraction of the energy in GP1, and correspondingly for transfers to successively higher trophic levels.

To summarize the results for terrestrial ecosystems, it seems safe to say that the efficiency at every step is roughly 10 percent or a bit more. If figure 11.3 were drawn to scale (it isn't) each GP box would be one-tenth the size of the one below it and ten times the size of the one above it. In marine ecosystems, efficiencies may be considerably higher, in some transfers possibly as high as 70 percent.[16]

Consider the efficiency with which trained, well-fed humans can perform athletic feats; the efficiency is said to be as high as 30 percent.[17] Admittedly, we are dealing here with a few high-quality specimens of one species rather than a whole trophic level; the athletes tested would belong to trophic levels 1 and 2 if they were meat-and-vegetable feeders, or to level 1 if they were strict vegans. In any case, humans at their best appear to take up energy more efficiently than the average for other terrestrial organisms, but much less efficiently than some marine organisms.

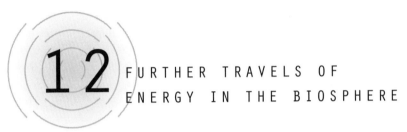

# 12 FURTHER TRAVELS OF ENERGY IN THE BIOSPHERE

*Captured Solar Energy: Its Final Destination*

As we saw in chapter 11, nearly 2 billion billion ($2 \times 10^{18}$) kJ of solar energy are captured each year by photosynthesizing plants.

What becomes of all the captured energy? Some of it travels from plants to herbivores and on to carnivores, up a number of familiar food chains, and is partly dissipated by respiration at each level. But what happens to the rest—the energy still unused when the animals at any level of a food chain die?

The bulk of the captured energy, in any case, never gets beyond the lowest level of a food chain: enormous quantities of vegetation die without being eaten. Where, then, does all the energy in the dead vegetation go?

The answer to both questions is the same: the energy trapped in dead organisms, whatever their trophic level, is ultimately released by decay or by burning. One or the other—decay (more formally, *decomposition*) or burning (*combustion*)—terminates the temporary existence of the organic molecules that living matter consists of: the organic molecules are broken down to inorganic ones, preponderantly carbon dioxide and water, and the chemical energy stored in them is liberated. At the same time, oxygen is used up. In other words, the photosynthetic reaction described in chapter 12 happens in reverse, thus:

$$CH_2O + O_2 \Rightarrow CO_2 + H_2O + energy.$$

Here $CH_2O$ is shorthand for carbohydrates in general and represents the organic molecule being broken down.[1] With inconsequential changes, the formula could be adapted to describe the breakdown of any organic molecule.

The important point is that the quantity of energy liberated when organic matter is destroyed is always *exactly* equal to the quantity of solar energy used to create it. This is true irrespective of whether combustion or decomposition brings about the disintegration. Combustion, as in a forest fire, usually happens fast, so that a large proportion of the energy released is free energy (see chapter 10). Decomposition is slow; much of the energy released is at a low temperature—it is entropy. We must next consider what causes decomposition: How does it happen?

*Detritus Food Chains*

Decomposition is the consumption of dead plant and animal matter by the many kinds of bacteria and molds that feed on it; these bacteria and molds, known collectively as *decomposers,* use dead organic material as *their* energy source. Decomposers, as a group, belong to food chains of their own, distinct from (though linked to) the familiar plants-herbivores-carnivores food chains we considered in chapter 11. The decomposers' food chains are known as *detritus food chains.* Like "ordinary" food chains, they transfer chemical energy from one group of organisms to another in a series of steps. Although the organisms concerned are for the most part microscopic, the energy they transfer is considerable.

Large amounts of detritus—dead organic material—can be found lying around nearly everywhere you look (the cautious "nearly" is to exclude deserts, and also concrete and other synthetic surfaces). Its presence is too obvious for comment; it attracts attention only when we have to rake fallen leaves or scramble over downed trees.

The bulk of detritus consists of dead plant material. If we exclude vegetation that happens to be consumed by fire, about 90 percent of all plant material becomes detritus;[2] either the plants die or else they are eaten and excreted undigested in herbivores' feces. All dead, uneaten animal remains become detritus too. It is the material at the bottom of every detritus food chain, and its energy, stored as the chemical energy in organic molecules, is ultimately derived from the sun. In terms of trophic levels, detritus is to a detritus food chain what living vegetation is to an "ordinary" food chain.

Some of the energy in detritus is dissipated in the life processes of the bacteria and molds that consume it. But a lot of the energy is passed on when the decomposers themselves are eaten by larger organisms, known as *detritivores*. Typical detritivores are beetle grubs feeding on rotting (that is, decomposing) wood, maggots feeding on rotting meat, immature aquatic insects feeding on "drowned" leaves at the bottom of still water, and clams feeding on detritus particles in salt water. The detritivores appear to be eating dead food; in fact, they are eating the decomposers together with some of the incompletely rotted organic material softened by the decomposers. Many of the detritivores are themselves eaten, perhaps by insect-eating birds, perhaps accidentally by grazing mammals. When this happens the energy is transferred into an ordinary, macroscopic food chain at one of the carnivore levels. The matter containing the energy becomes flesh, and in due course, dead flesh; that is, it becomes detritus again, at the starting point of another detritus food chain. Any given organic molecule may cycle through macroscopic food chains and detritus food chains alternately, over and over again, before ultimately disintegrating to carbon dioxide and water.

All these seemingly negligible natural events are not negligible at all when taken together as transfers of energy. Bacteria "consume almost everything in their environment . . . [and] reproduce more rapidly than other living organisms." This makes them "a major source of energy for other consumers."[3] If we disregard fires, then bacteria and molds are the final dissipaters of captured solar energy. The energy has to go somewhere, and only a minute fraction of it goes into long-term storage in fossil fuels—coal and oil—and into short-term storage as peat.

Detritus food chains are as important in the ocean as on land. Huge numbers of plankters—both green phytoplankters and the zooplankters that feed on them—escape being eaten by larger organisms and simply die; the result is a rain of tiny bodies sinking slowly through ocean waters nearly everywhere. They are the detritus of the ocean, and they form the base of marine detritus food chains.

The chemical energy in an organic molecule lasts for as long as the molecule lasts, from its creation to its disintegration. The lifetime of a carbohydrate molecule, for instance, starts with its "birth," energized by sunlight, and ends with its "death" by bacterial decomposition or fire; its lifetime may be a few seconds or millions of years. Consider the extremes. When fire destroys growing grass, some of the energy liberated must have been captured only seconds before. When bacteria decompose the last sliver of wood from what was once an ancient rain-forest tree, the final molecules to disintegrate may have existed for thousands of years, first while the tree was alive and then as it gradually decayed.[4] The tiny fraction of detritus fossilized to form coal and oil can persist with its energy intact for millions of years; some of the energy is dissipated when the fuel is burned by humans; as for the rest, all we can say is that it won't outlast the planet.

## Energy without Sunlight

As we've noted before, *nearly* all living things obtain their energy, directly or indirectly, from the sun. Here we consider the organisms that the word *nearly* excludes. They are bacteria that obtain their energy as chemical energy, direct from inorganic molecules—from molecules of mineral origin. They don't need the energy of light, which means they can function in the dark. The process is called *chemosynthesis* to differentiate it from photosynthesis.[5] Both kinds of synthesis require two things—a source of energy and a source of carbon. The difference between them is in the energy source; the carbon source is the same, carbon dioxide in air or carbonate in water. Both kinds of synthesis "fix" carbon.

The discovery that organic molecules—the molecules living organisms are made of—could be synthesized in the absence of chlorophyll, by a process other than photosynthesis, was one of the giant steps in nineteenth-century science. It was made by the Russian biochemist Vinogradsky, who, inexplicably, has not achieved the widespread fame he deserves.[6]

He wrote, "[Chemosynthesis] is contradictory to that fundamental doctrine of physiology which states that a complete synthesis of organic matter cannot take place in nature except through chlorophyll-containing plants by the action of light."[7] He discovered that, on the contrary, some bacteria can gain the energy they need for growth and reproduction by oxidizing inorganic compounds.

One species of bacteria that does this is *Thiobacillus oxidans;* members of the species are the bacteria responsible for acid mine drainage. They oxidize sulfur compounds such as the mineral pyrite (otherwise known as iron sulfide or fool's gold) to sulfuric acid and in so doing liberate the energy they require to synthesize the organic molecules necessary for life and growth. Another sulfur compound they often oxidize is the "rotten egg" gas, hydrogen sulfide. *Thiobacillus oxidans* and other so-called iron bacteria also oxidize iron; the re-action they cause is the same as rusting, which, as we saw in chapter 10, is an exothermic (energy yielding) reaction. Various species of iron bacteria live in the water of acid bogs, where they oxidize dissolved forms of iron, leaving the rust-colored coating (indeed, it *is* rust) often found on the bottom of bog pools when they dry up.[8]

Another group of chemosynthetic bacteria are the *nitrifying bacteria* that live in the soil. Some species obtain their energy by oxidizing ammonia to ni-trite and others by oxidizing nitrite to nitrate.[9]

Note that chemosynthesizing bacteria use the chemical energy they cap-ture in the same way that photosynthesizing plants use the light energy *they* capture. In both cases the energy is used to fix carbon and create organic mol-ecules. Then, while they live, the bacteria (like the plants) break down some of the organic molecules to release energy for life processes. Some bacteria (not all) do this by ordinary respiration.

Chemosynthetic reactions are crucial to the growth of plants; they ensure that plants get the nitrogen they must have in a form they can use. The en-ergy transfers involved are small, however: chemosynthesis is relatively unimportant from the energy point of view—on our planet at any rate. This is because the bacteria concerned (apart from those that live in soil) tend to oc-cupy uncommon, sulfur-rich habitats in pitch darkness. Some examples: "cold seeps" on the seafloor, where cold water with sulfides and other chemicals dis-solved in it seeps up at depths where sunlight can't penetrate;[10] oil seeps on the sea floor, also at dark depths;[11] and the hot, sulfurous waters of deep-sea hydrothermal vents.[12]

The characteristic of chemosynthesis that makes it noteworthy in the con-text of "bioenergy" is that, as Vinogradsky proved, it does not come from sun-light. Chemosynthesis could conceivably be the fuel of life on planets not con-stantly bathed in strong sunlight as the earth is. To regard a habitat as "unusual" because it is poorly represented on earth is a geocentric prejudice.

*Rocks Built by Sunlight*

As we have just seen, certain bacteria use minerals as fuel. In striking contrast are other tiny organisms that use solar energy to build rocks. Biological "rock

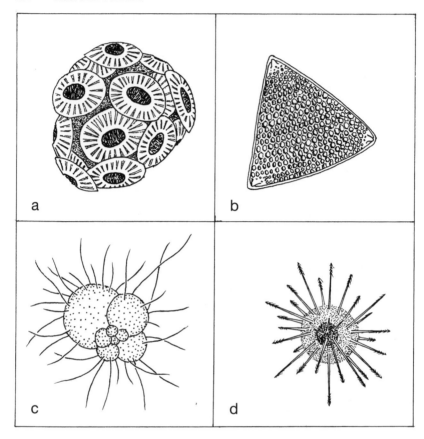

Figure 12.1. Four microscopic organisms whose hard parts accumulate to form rocks: (a) the coccolithophore *Emiliana huxleyi* (the covering scales are coccoliths); (b) a diatom; (c) a foram; (d) a radiolarian.

building," the topic we come to now, is the direct opposite of chemosynthesis: whereas chemosynthetic bacteria use rocks to build life, rock-building organisms use the products of photosynthesis to build rocks; the rocks are known as *biominerals*.

Familiar examples are coral reefs and those limestones that consist of the shells or skeletons of billions of microscopic organisms. Less abundant are flints and cherts, rocks formed from the silica skeletons of microscopic diatoms and radiolarians (fig. 12.1).

First consider coral. The individual organisms in living coral are tiny polyps—miniature versions of sea anemones, to which they are closely related. The polyps secrete the mineral calcite (limestone) and use it to construct cup-shaped external skeletons for themselves. Usually the polyps live in closely packed colonies, with the walls of the cups fused together. When the polyps are dead, the solid masses of calcite skeletons are coral rock—a biomineral.

While they are alive the polyps feed like sea anemones, grasping their prey—mostly plankton animals and minute crustaceans—with their tentacles. Nearly all the coral living in shallow seas where sunlight penetrates have plantlike, photosynthesizing green cells living inside the polyps. The green cells are not part of the polyps; they are separate organisms living in a symbiotic partnership that benefits both. The green cells benefit by being protected from predators, and the polyps benefit because the green cells' photosynthesis absorbs and disposes of surplus carbon dioxide.

Next consider chalk, another form of limestone consisting of the skeletons of other tiny marine organisms. The skeletons collect on the sea floor as *calcareous ooze*, which, if it is free of sand, eventually hardens to chalk, a soft, pure white rock. Chalk occurs as long ranges of hills in southeastern England and northeastern France, and where the hills have been cut through by the English Channel, the chalk is exposed as the famous white cliffs of Dover. Chunks of flint, formed from the siliceous skeletons of diatoms and radiolarians, can be found embedded in the chalk.

The most vigorous calcite producers are certain microsocopic phytoplankters known as *coccolithophores*. They secrete *coccoliths*, tiny, intricately patterned calcite "scales" that coat the outside of each phytoplankter's cell wall, presumably functioning as protective armor (see fig. 12.1). After the living cells that bore them die, the coccoliths are all that remain, and they are often the most abundant ingredients in calcareous ooze. Other organisms that contribute to the ooze are foraminifera (forams for short), and the rudimentary shells of "sea butterflies" or pteropods, small mollusks often found in great numbers in the surface waters of the oceans. (Note that calcareous ooze can accumulate only at depths less than 4.5 km; below that level, calcite dissolves.)

The link between solar energy and chalk has been most clearly revealed in studies of living coccolithophores of the species *Emiliana huxleyi*, known to oceanographers everywhere as *Ehux*.[13] As members of the plankton, they float in the uppermost layer of the ocean, where the sunlight they need for photosynthesis can reach them. They sometimes occur in immense "blooms," the consequence of population explosions. A bloom colors the sea turquoise in huge, irregular patches; the patches may be larger in area than 100,000 square kilometers (the size of England), and they show up conspicuously in photos taken from space.

The way photosynthesis—hence solar energy—affects coccolith formation has been discovered by observing the process while it is happening.[14] When calcite crystallizes in the absence of *Ehux*, the crystals are plain rhomboids, independent of one another. But when calcite crystallizes inside an *Ehux* cell, the process is controlled by giant carbohydrate (polysaccharide) molecules, produced by photosynthesis; the particular polysaccharide involved is structurally "one of the most complicated ever described,"[15] and by binding to the growing calcite crystals it controls the pattern they construct. The result is the formation of coccoliths, each a symmetrically patterned, perforated structure only a few micrometers in diameter. The buoyant *Ehux* cells bearing them are eaten by copepods—minute crustaceans in the plankton—and excreted in the copepods' feces, which sink to the bottom, where they carpet the seafloor over immense areas. In time (millions of years), these carpets of calcite ooze will undoubtedly become chalk cliffs.

To summarize: A variety of rocks are biominerals, consisting of the skeletal remains of microscopic organisms. Four groups of these organisms are especially noteworthy—coccolithophores, forams, diatoms, and radiolarians. Of these, the first two secrete calcite (which forms ordinary limestone and chalk) and the second  two secrete silica (which forms chert and flint). Two of the groups (coccolithophores and diatoms) contain chlorophyll, enabling them to photosynthesize; the other two (forams and radiolarians) are animals, somewhat like amoebas with external "skeletons," which obtain their solar energy at second hand. In all four groups, the hard parts are intricately structured and sculpted into astonishingly beautiful forms (see fig. 12.1); they have been called "the miniature jewelry of the abyss."[16]

Energy, originally from the sun, goes into the construction of these "jewels" and is stored in their elaborate geometrical structures. When the structures are crushed, the structural energy is dissipated. This is a microscopic version of what happens when big rocks are broken by weathering, as we saw in chapter 9. Many of the microscopic skeletons remain undamaged or only slightly damaged within the rocks; it is difficult to foresee how the energy they still hold will ultimately be dissipated.

# 13 THE WARMTH OF THE EARTH
## NUCLEAR REACTIONS SUSTAIN ALL LIFE

*Heat from Solar and Terrestrial Sources*

Up to this point we have considered energy originating in the sun. Think of the atmosphere: all the winds, from light airs to hurricanes, are energized by the sun's heat. Think of the oceans: ocean currents wouldn't flow were it not for the sun's heat. Think of living things: almost all depend on the energy of sunlight captured by photosynthesis (the exceptions are some species of bacteria mentioned in chapter 12). Heat from the sun causes surface water to evaporate, giving rise to clouds yielding rain and snow, which nourish rivers and lakes. Flowing water, aided by wind and wind-driven waves, causes erosion, which shapes the face of the earth.

Not all the earth's energy comes from sunlight, however. A small fraction—one part in four or five thousand—comes from the earth's internal heat. This is the energy that shifts tectonic plates and that powers earthquakes and volcanoes. It heats rocks to temperatures high enough to change them to metamorphic rocks—limestone to marble, for example, and granite to schist.

It drives the circulation of liquid iron in the earth's core, which makes the whole earth a magnet. It heats hot springs and geysers.

If the sun were suddenly to disappear, the atmosphere and oceans would become silent and still; the oceans would freeze solid, except where submarine hot springs (*hydrothermal vents*) emerge in the depths; every living thing on the planet's surface would die. At the same time, hot springs would continue to bubble up from the earth's interior; tectonic plates would continue to drift, volcanoes to erupt, and earthquakes to shake our planet. Rocks would still be metamorphosed, and the earth would still be a magnet.

All these energetic events would keep happening even though the energy from inside the earth is, as noted above, four or five thousand times less than the solar energy coming to us from outside. Both kinds of energy eventually leave the earth by radiation skyward: earth's heat does not accumulate. Solar heat is radiated back into space, while the internal heat, what little there is of it, escapes through the surface and is gone forever.

Let's see how these two radiation rates, measured in watts per square meter (W m$^{-2}$) of radiating surface, compare with a couple of other, easily visualized radiation rates, those of a 100-watt lightbulb and a clothed human body. We now have four radiation rates to consider, measured at the surfaces of the objects concerned. Listing them from greatest to least, they are: for the light and heat from the 100-watt lightbulb, about 2,000 W m$^{-2}$; for the sunlight reradiated from the earth, 340 W m$^{-2}$; for the internal heat radiated from the earth, 0.08 W m$^{-2}$; and for the warmth radiated from a clothed person on a very cold, windy winter day,[1] 0.002 W m$^{-2}$.

Note how the radiation rates of the small objects (the lightbulb and the person) bracket the rates at which the earth radiates its two kinds of energy (external and internal). The surface of a 100-watt lightbulb is obviously much hotter, and is radiating much faster, than the surface of the ground. Also (although this is less obvious), the clothes of a person exposed to a bitter winter wind are colder, and are radiating more slowly, than the surface of the ground. In short, if you dress warmly and keep your clothes on, you will lose your warmth more slowly than the earth loses its warmth.

*Atomic Nuclei: The Source of All Energy Heats the Earth*

We now inquire how the sun's heat and the earth's internal heat come into existence. Nuclear reactions are the most important cause, *nuclear fusion* in the case of the sun, and *radioactivity* of a type that can, broadly speaking, be called *nuclear fission* in the case of the earth. In fact "the energy involved in almost all natural processes can be traced to nuclear reactions and transformations."[2] Fusion is the principal source of the sun's heat, and fission is the principal source of the earth's; these are the heat sources we consider in this chapter. Both the earth and the sun also have another supply of heat: the heat remaining from the time of their formation about 4.5 billion years ago, some of which still remains (see chapter 14).

A digression on the basics of atomic structure is necessary here. As is well known, an atom consists of an exceedingly small nucleus surrounded by a "swarm" of even smaller electrons, moving in a comparatively large space centered on the nucleus. Each electron has a negative electrical charge, and the nucleus contains an equal number of positively charged particles (protons), making the whole atom electrically neutral. Every chemical element is distinguished from all the others by the number of electrons it has. An atom of hydrogen, the lightest element, has a single electron; an atom of uranium, the heaviest naturally occurring element, has ninety-two (still heavier elements, with more electrons, have been created artificially).

So far, so good. We are about to consider events in the nuclei of atoms, which, as we shall see, are many orders of magnitude more energetic than chemical reactions of the kind considered in chapter 10. Those reactions—photosynthesis and combustion, for instance—involve only the electron swarms of the participating atoms: the atomic nuclei take no part.

The tremendous energy contrast between ordinary chemical reactions and nuclear reactions cannot be overemphasized. The contrast becomes somewhat easier to appreciate when you compare the relative sizes of atoms and nuclei. Most of the volume of an atom is the space occupied by the electrons, so most of an atom's volume is empty space. The volume of a carbon atom, for instance, is about $2.5 \times 10^{-24}$ ml (one milliliter is about the volume of a sugar cube). If the carbon atom were represented by a globe the size of the earth, the nucleus would be a ball at the center with diameter less than 100 m.

The contrast in sizes shows that there must obviously be a corresponding contrast in densities. The density of a solid object such as a pebble is its mass divided by its volume, with the measured volume including all the empty space in every atom. Nuclear material lacks empty space, making its density approximately one hundred trillion times greater. In fact, nuclear density is about a quarter of a billion metric tons per milliliter (more precisely, $2.4 \times 10^{17}$ kg m$^{-3}$).[3]

The inconceivably high density of an atom's nucleus makes it intuitively reasonable to suppose that the nucleus is held in one piece—or, to look at it the other way round, prevented from flying apart—by unimaginably strong forces. Intuition is correct: the bonds holding an atom's nucleus together are tens or hundred of millions times stronger than the chemical bonds described in chapter 10, which hold the atoms in a molecule together. This pent-up force is potential energy waiting to be liberated. As we have noted already and will now consider in more detail, the liberation is brought about by nuclear fission in the earth's interior and by nuclear fusion in the sun. The energy is known as the *binding energy of the nucleus*.

## The Binding Together of Nuclear Particles

Before going further, we must consider what it is that is bound: What kind of elementary particles are held together in an atom's nucleus? The answer is protons and neutrons. Each has a mass close to two thousand times the mass of an electron.[4] Every proton carries a single positive charge of electricity, exactly balancing the negative charge of an electron; neutrons are electrically neutral. Protons and neutrons together, known jointly as *nucleons*, are the particles that make up nuclei.

The number of protons in the nucleus of any given element is the same as the number of electrons swarming around the nucleus, as we noted above. The number of neutrons varies, from zero in a  common hydrogen atom (whose nucleus consists of a single proton) to 146 in the commonest form of uranium. In the nuclei of light elements, such as calcium and carbon, the protons and neutrons in the nucleus are equal in number, but in heavier elements neutrons outnumber protons; in the uranium atom, the number of neutrons is more than 50 percent greater than the number of protons.

It often happens that not all the atoms of a given element have the same number of neutrons in the nucleus, even though they all have the same numbers of protons. For example, the number of protons in the nucleus of every atom of chlorine is 17, but the number of neutrons is *not* the same in every nucleus: about 75 percent of them have 18 neutrons, for a total of 35 nucleons; the rest have 20 neutrons, for a total of 37 nucleons. These are the two *isotopes* of chlorine, known respectively as chlorine-35 and chlorine-37 (or by the symbols $^{35}Cl$ and $^{37}Cl$). The isotopes of uranium are known by name to everybody who has read about atomic bombs and atomic energy; the familiar ones are uranium-235 (with 235 nucleons comprising 92 protons and 143 neutrons) and the far more abundant uranium-238 (with 92 protons and 146 neutrons). Carbon has three isotopes: about 99 percent of all carbon is carbon-12,

1 percent is carbon-13, and a very tiny fraction, far too small to be recorded as a percentage, is carbon-14. Their nuclei all have 6 protons and 6, 7, or 8 neutrons respectively.

This digression on isotopes is a necessary preliminary to the statement that the binding energy of a nucleus depends on the number of nucleons it contains. The strongest of all nuclei—those with the greatest binding energy per nucleon—are those of iron-56, which have 26 protons and 30 neutrons. They are the most stable nuclei, more stable than either lighter ones or heavier ones. The reason a nucleus of intermediate size is held together more strongly than smaller or larger nuclei will become clear in a moment, when we consider the nature of the force holding nucleons together. The force does not act in the same way as the force governing the behavior of larger objects when they are electrically charged.

As everybody who has been infuriated by static cling knows, two electrical charges of the same sign—two positive charges or two negative charges—repel each other, whereas two unlike charges—one positive and one negative—attract each other. This leads one to expect that an atom's nucleus would automatically fly apart, because all its charged particles are positively charged protons. It doesn't because it is held together by a vastly more powerful force, the *strong interaction* between elementary particles. This strong force holding particles together overwhelms the much weaker electrical force that drives similarly charged particles apart; but it acts only over short distances—exceedingly short distances, a trillionth of a millimeter ($10^{-15}$ m) or less. If a nucleus is smaller than this in diameter, then every one of its particles is attracted to every other by the strong interaction; therefore the more numerous the particles and, consequently, the more numerous the pairwise strong interactions, the more strongly the nucleus as a whole is held together. In brief, the bigger the stronger, but only up to a limit. If a nucleus consists of so many nucleons that its diameter exceeds the range of the strong interaction, some of its protons will be spaced far enough apart to repel each other by electrical force. The largest nucleus in which every nucleon is attracted to every other by the strong interaction is (as you will have guessed) iron-56. Nuclei larger than this are increasingly unstable. The heaviest nucleus to occur naturally is uranium-238.

*Nuclear Fusion: $E = mc^2$*

We have reached the stage (at last) where we can describe nuclear fusion and (in the following section) nuclear fission and comprehend the source of the en-

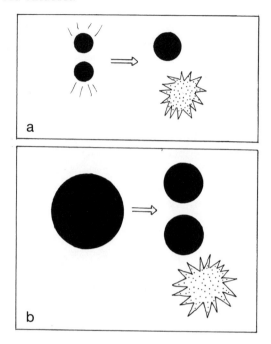

Figure 13.1 (a) Nuclear fusion. (b) Nuclear fission. Masses are shown black, and energy is shown stippled. In each figure the total mass to the left of the arrow, showing the reactant(s), exceeds the total amount to the right, showing the product. The excess mass is liberated as energy. In both cases the loss of mass is exaggerated for clarity.

ergy these two nuclear events liberate. What happens is shown diagrammatically in figure 13.1: the upper panel shows fusion, the lower panel fission.

Fusion occurs when two lightweight nuclei happen to come so close to each other that the strong interaction pulls them together, overcoming their tendency to repel each other because both are positively charged electrically. They combine to form a larger nucleus. The larger nucleus (the *product* nucleus) has a slightly smaller mass than that of the two nuclei (the *reactant* nuclei) that combined to form it. The seemingly vanished mass hasn't really vanished, however; it has been converted into energy, in accordance with Einstein's famous formula, $E = mc^2$. $E$ is the energy, measured in joules; $m$ is the mass, in kilograms, that has "gone missing"; and $c$ is the velocity of light, 300 million meters per second or, in scientific format, $3 \times 10^8$ m s$^{-1}$.

The fusion of a pair of nuclei of "heavy hydrogen" will serve as an example.[5] Heavy hydrogen (also called *deuterium*) is an isotope of ordinary hydrogen: whereas an atom of ordinary hydrogen has a single proton for its nucleus (its symbol is $^1$H), heavy hydrogen has a nucleus of two nucleons, a proton and a neutron (its symbol is $^2$H). When two of these heavy hydrogen nuclei collide, they combine to form a single nucleus of helium, which has four nucleons (two protons and two neutrons); its symbol is $^4$He. The combined mass of the two reactant nuclei is $6.68901 \times 10^{-27}$ kg, and the mass of the product nucleus (helium) is $6.64649 \times 10^{-27}$ kg.[6] The mass of the product is therefore $0.04252 \times 10^{-27}$ kg less than the mass of the reactants. This is the number to be substituted for $m$ in the "famous formula"; $c^2$ is $9 \times 10^{16}$ m$^2$ s$^{-2}$. It takes only a pocket calculator to confirm that $E = mc^2 = 3.827 \times 10^{-12}$ J. Finally, we convert joules to electronvolts (1 J = $6.24 \times 10^{18}$ eV; see chapter 10), as the more convenient units for measuring such minuscule amounts of energy. The answer is $E = 24$ million electron volts, written 24 MeV, per nucleus of helium.

This is the energy liberated when two heavy hydrogen nuclei combine to form a helium nucleus, and it is also the bonding energy of the helium nucleus. To split a helium nucleus into two heavy hydrogen nuclei, you would have to supply 24 MeV to get the job done. The 24 MeV you provide would become converted into mass—the mass by which two heavy hydrogen nuclei exceed the mass of a single helium nucleus. Whichever way the reaction goes, fusion or fission, the mass-plus-energy (or mass-energy for short) is the same at the end of the reaction as it was at the beginning: it is conserved. This is an informal statement of the law of the conservation of mass-energy, which replaced two famous nineteenth-century laws: the law of the conservation of mass and the law of the conservation of energy.

Note that the bonding energy of a helium nucleus, namely 24 MeV, is more than 5 million times the energy needed to separate the two hydrogen atoms forming a hydrogen molecule, which, as we saw in chapter 10, is 4.5 eV. This is a good example of the tremendous difference between the energy holding a nucleus together and the energy holding the atoms in a molecule together.

On a more "human" scale, we can say that the fusion energy obtainable from 150 milligrams of heavy hydrogen (picture a 500 milligram vitamin C tablet for comparison) is about the same as the combustion energy produced by burning 2,700 liters of gasoline.

Nuclear fusions do not happen naturally on earth (their unnatural occurrence is considered in chapter 19). This is because natural conditions on earth never allow a pair of nuclei to come close enough to each other for the strong interaction force to take hold and drive them together. Collisions between nuclei happen only in environments where the temperature and pressure are, literally, out of this world; they happen in the interiors of stars, including the sun. The fusion of pairs of heavy hydrogen nuclei to form helium nuclei is, indeed, the sun's chief source of energy. The whole reaction has a few more steps than the simple fusion described in detail above, because the sun's "ordinary" hydrogen must first be converted to heavy hydrogen. The principle is the same, however, and the energy liberated per helium nucleus formed is a little more: it is 24.7 MeV.[7]

Bear in mind that though nuclear fusion never happens naturally on earth, it provides practically all the energy we have. It happens in the sun. "Homegrown," locally produced energy amounts to only one part in four or five thousand of the total energy that keeps the earth going, as we noted at the beginning of this chapter. Most of it is produced by nuclear fission.

## Nuclear Fission: $E = mc^2$ Again

Nuclear fission takes place when a heavy nucleus splits into two or more product nuclei. The combined mass of the product nuclei falls short of the mass of the reactant nucleus (the one that split), and the vanished mass instantly becomes energy.

As an example, let's compute the energy liberated when a nucleus of uranium-235 splits into one nucleus of strontium-90 and one nucleus of cerium-144 plus a neutron (there are also four leftover electrons).[8] The mass that "disappears" amounts to $0.3561 \times 10^{-27}$ kg. Applying the formula $E = mc^2$ shows that this mass becomes $32.05 \times 10^{-12}$ J, or about 200 MeV. This is the energy liberated by the fission of one uranium-235 nucleus.

The arithmetic is straightforward, but where and why does nuclear fission happen? The question has two answers: First, it happens naturally and spontaneously, in radioactive elements contained in the rocks the earth is made of. This naturally occurring nuclear fission is what maintains the warmth of the earth's interior, keeping the tectonic plates in motion, causing mountains to rise up, and driving a variety of other natural processes.

Second, it can be made to happen, unnaturally fast, by technological means. "Atomic" bombs, of the kind first used in war in 1945, get their energy from nuclear fission that is caused to happen in a confined space and at a tremendously rapid rate. (In contrast, "nuclear" or "hydrogen" bombs are powered by nuclear fusion.) Controlled nuclear fission, proceeding at a more leisurely pace, is the energy source in current atomic power stations.

*Un*natural nuclear fission is discussed in chapter 19. Here we consider how natural nuclear fission warms the earth. The nuclei of several elements are involved.[9] The fissions responsible differ from the example already described in that the nuclei resulting from the split are very unequal in size. In the fission we've already considered, a nucleus with 235 nucleons split into nuclei with 90 and 144 nucleons plus one "spare" neutron. By contrast, in the fissions whose energy warms the earth, a heavy nucleus splits unequally: one of the product nuclei is a helium nucleus with 4 nucleons (2 protons and 2 neutrons), and the other is a nucleus only 4 units lighter than the original.

Fission of this kind—the splitting off of a lightweight helium nucleus from a heavy nucleus—is known as *radioactive α-decay* (α is the Greek letter *alpha*); it is one kind of radioactivity. There are other kinds as well, namely β-decay and γ-decay, but they add negligibly to the earth's interior heat, the topic that concerns us here. Radioactive α-decay isn't a typical example of nuclear fission because the products of the fission are so unequal in size. It is called α-decay because radioactivity was discovered, and its different forms named, years before it became clear that the α-particles emitted in radioactive α-decay are identical with the nuclei of helium atoms. Indeed, the terms "α-particle" and "helium nucleus" mean the same thing.

The nuclei whose splitting contributes most to the earth's heat are two isotopes of uranium (uranium-238 and uranium-235) and one of thorium (thorium-232). A single nucleus of any of these splits repeatedly, hiving off a helium nucleus and releasing energy at each split, before winding up as a stable nucleus immune to further splitting. Electrons in the space controlled by the nucleus are also lost, so that the final, stable nucleus differs chemically from its "ancestor" nucleus as well as having fewer nucleons.

Thus a nucleus of uranium-238 splits off 8 helium nuclei in succession, for a loss of 32 nucleons in all; it also changes chemically, becoming an isotope of lead, namely lead-206. In similar fashion, a nucleus of uranium-235 loses a total of 7 helium nuclei before reaching stability as a different isotope of lead, this time lead-207. Thorium-232 loses 6 helium nuclei and becomes lead-208.

The time intervals between successive fissions are a matter of chance; they are intrinsically unpredictable. Some of the nuclei persist for millennia, others for less than a second.

These three "decay" processes provide most of the earth's internal heating at present. The three "fuels" (uranium-238, uranium-235, and thorium-232) are, of course, in the process of being consumed—used up—as surely as, though more slowly than, fossil fuels like coal and oil are being used up by humankind. What's more, the natural process is unstoppable. It also follows that the quantities and the relative proportions of the different radioactive elements in the earth are continuously changing through geologic time; some that were present in the distant past have already been used up. More on the subject in chapter 14.

In the meantime, it's worth reemphasizing that practically all the energy available on earth comes from nuclear reactions of one kind or another. Nuclear fusions in the distant sun are the most lavish source. Nuclear fissions, deep underground where the sun's warmth cannot penetrate, continuously heat the earth's interior and energize the many processes always going on there.

# 14 THE EARTH'S INTERNAL ENERGY

*The Earth's Layered Structure*

The interior of our earth is full of heat, noise, and movement—in a word, it is brimming with energy. To understand what is happening, we must consider the earth's structure; it consists of several layers in the form of spherical shells surrounding a core; the layers differ from each other chemically and physically.

Figure 14.1a shows a section of how the earth would appear if it were sliced open through the center; figure 14.1b is an enlargement of the outermost part showing more detail. Each layer is named, and as the caption explains, the shading shows whether it is solid, liquid, or ductile (plastic). A ductile material behaves like glacier ice or like glass: it breaks cleanly in response to an abrupt force, but when acted upon by a long-lasting, slow-acting force it gradually becomes distorted so that it "creeps," or flows, exceedingly slowly. Figure 14.1a also shows the temperature at various depths. The information in the diagrams can be absorbed at a glance, but discovering it has taken years of research. Much remains to be learned, and investigations con-

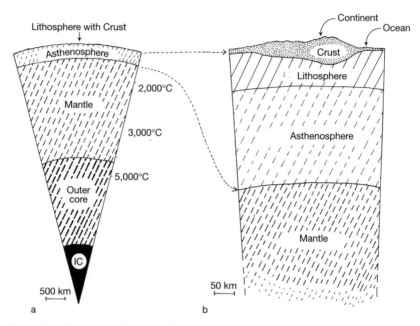

Figure 14.1 (a) A sector of the earth in cross section, showing the layers described in the text; IC is the inner core. Note the temperatures at different depths. (b) The outer part of the section in (a) enlarged ten times to show details of the crust and the rest of the lithosphere. Liquid and plastic (ductile) layers are hatched with broken lines. Note: The crust is classified as part of the lithosphere and the asthenosphere as part of the mantle.

tinue: new discoveries are made frequently and new data recorded; methods of observation are being refined, and theories modified. What we are about to consider is very much work in progress.

The knowledge figure 14.1 is based on has, for the most part, been obtained by recording the speed at which seismic waves pass through the earth for various distances. Sometimes the seismic waves caused by earthquakes are used, and sometimes (since the 1960s) the waves caused by nuclear explosions. The time it takes for seismic waves to travel from their point of origin to the point where they are recorded depends, obviously, on the distance they travel and also on the temperature of the rock they travel through. Waves travel more slowly through hot, soft rock and more quickly through comparatively cool, hard rock. The waves seldom travel in straight lines; they are variously re-

THE EARTH'S INTERNAL ENERGY     151

flected and refracted whenever they encounter temperature changes. Not surprisingly, data interpretation is difficult.

Let's assume that all the difficulties have been solved and take a brief look at the current state of knowledge, realizing that revisions and extensions are coming all the time. Then we can ask, and try to answer, how and where energy is generated and (in the succeeding chapter) how it is ultimately dissipated. Generation and dissipation are equally important.

Now for the earth's structure: simplifying matters considerably, the layers, from the surface downward, are as follows.

The outermost layer, which is not uniform, is the *crust:* oceanic crust, under the ocean floor, is a thin layer of relatively heavy rocks—chiefly basalt—whereas continental crust, which underlies continents and continental shelves, is a thicker layer of less dense rock—chiefly granite and related rocks.

Below the crust is the *lithosphere.* This is a layer of brittle rock broken into detached pieces, the *tectonic plates;* they slowly drift this way and that over the earth's surface. When they come together, the rim of one rides up over the other; the lower plate is forced down to great depths—it is said to *subduct*—while the leading edge of the upper plate scrapes off any sediments it encounters. The continental crust can be fairly regarded as the scum of the earth. Indeed, it is usually classified as merely the outermost part of the lithosphere.

Below the lithosphere is the *asthenosphere,* a layer of soft, partly molten rock. It acts as a lubricating layer, enabling the tectonic plates above it to slide, or "drift," over the solid layer below it; there is more to be said about plate drift in a subsequent section.

The next layer is the *mantle.* It is a ductile solid, consisting of silicate rock. The asthenosphere, being merely the  soft outermost skin of the mantle proper, is classified as part of the mantle in the same way as the crust is classified as part of the lithosphere.

Below the mantle is the *core* of the earth. It has two layers, an outer liquid layer encasing an inner solid layer.

All the way from the surface to the center, 6,370 km down, both the temperature and the pressure rise, abruptly at some levels and more slowly at others. Rising temperature melts rock *if* the pressure doesn't increase; conversely, increasing pressure solidifies molten rock *if* the temperature doesn't rise. When temperature and pressure rise together, the dominant factor alternates, which explains why hard, solid layers alternate with ductile or liquid layers. The layers differ from each other chemically, too. The mantle is made up mostly of silicates and the core of iron, which is much heavier.

So much for dry facts, which give no mental image of what the underground world is really like. Indeed, the human mind is incapable of envisioning it—not that vision, in the ordinary sense, would be any help, because the darkness is absolute and the heat and pressure are lethal. This is not to say that conditions are uniform, however. Far from it. The contrast in density across the interface between the mantle and the core, nearly 3,000 km beneath us, far exceeds the contrast across the interface between the continental crust and the atmosphere. The latter interface is, of course, the surface we live on; it is "our" surface, the very ground we walk on, build houses and roads on, and drive autos on. From a geophysicist's point of view it is a much less remarkable surface than the one separating the mantle from the core.[1]

Across "our" surface, the difference in weight between a cubic meter of material above it (air) and a cubic meter below it (rock) averages 2,700 kg. Across the surface of the core, the difference in weight between a cubic meter of material above it (mantle rock) and a cubic meter below it (liquid iron) averages 4,330 kg. The latter pair of densities are current estimates, arrived at very indirectly. Not surprisingly, it's impossible to measure the density of rocks 3,000 km deep in the earth; the values have to be deduced from evidence of various kinds: from observations of the speed and direction of seismic waves; from the results of experiments carried out on rock specimens at temperatures and pressures far higher than any to be found at the earth's surface outside a physics laboratory; and by analyzing these data mathematically.

Now imagine a visitor from outer space inspecting our planet, a more exotic visitor than the stereotypical little green man. Arriving with no preconceptions, and with senses wholly unlike ours, the visitor might reasonably regard the core of the earth as the solid planet, with the mantle and everything outside it forming a many-layered "atmosphere"; this would be as reasonable a separation as the one we are used to. The surface we live on seems to us the only one, but that is merely because we live on it. To repeat, it is a much less noteworthy surface than the core-mantle interface.

The "inner planet"—the core—is slightly bigger than Mars. Its surface is believed to be as topographically interesting as the surface in our sense of the word, namely, the surface of the land and the floor of the oceans. Like them, the surface of the core has high mountains and deep valleys. Moreover, conditions are constantly changing at this inner surface, just as they are constantly changing, because of geological forces, at our surface. The changes are slow in human terms at both surfaces, but human notions of speed are irrelevant.

Nearly all the material of the earth's interior, except (perhaps) the solid iron of the inner core, moves constantly. The tectonic plates of the lithosphere drift; the ductile material of the mantle creeps; and the liquid iron of the outer core flows.

All this movement consumes energy, and the earth possesses energy of several kinds—thermal energy, kinetic energy, and magnetic energy. We begin by considering the thermal energy—heat—recalling that it comes from two sources: as explained in chapter 13, radioactivity is the primary source. An important secondary source is the residual heat remaining from the time of the earth's creation.

## The Radioactive Heat Source

As we noted in chapter 13, the earth's principal source of heat is a rather special kind of nuclear fission, radioactive α-decay; the radioactive elements of greatest importance in this context are uranium-238, uranium-235, and thorium-232.

Radioactive materials are most highly concentrated in the crust, and their concentration in the continental crust is double that in the oceanic crust. On average, the rate at which energy is produced by the continental crust amounts to less than half a microwatt (millionth of a watt) per ton of rock:[2] natural radioactivity can hardly be called an intense energy source. All the same, without it the earth's interior would be frozen and inert. There would be no earthquakes and no volcanoes. The tectonic plates would have been motionless for hundreds of millions of years, and with no collisions between the plates, mountain building would have stopped: the Rocky Mountains, the Andes, the Himalayas, and the Alps would never have existed. (For more on mountain building, see chapter 15.)

Radioactive elements are hundreds of times more concentrated in the crust than in the mantle; even so, fifty times as much heat is produced in the mantle as in the crust, because of the mantle's much greater volume. Although radioactive elements are more thinly spread below the crust than within it, the temperature continues to rise as the depth below the surface increases. But it rises much more rapidly in some layers than in others. Its rise is most rapid going down through the crust, and again 3,000 km below, across the boundary between the mantle and the core; this is the boundary of the "internal planet," where density increases so spectacularly, as we saw earlier. Between the *thermal boundary layers*, the temperature rises more gradually.

The foregoing account relates to present conditions. As we saw in chapter

13, the concentrations of the radioactive elements in the earth's interior are forever changing. Before examining the consequences, we should investigate the reason for this perpetual change.

Recall that an $\alpha$-decay event (the emission of an $\alpha$-particle) in a radioactive material is a chance occurrence, governed by probability. This makes it impossible to predict precisely when an individual radioactive nucleus will decay, but the probability that it will do so in, let us say, the next minute, or the next hour, or the next year, can be specified. The probability differs from one radioactive element to another: for instance, the probability that a uranium-235 nucleus will decay in a given time interval is far greater than the probability that a uranium-238 nucleus will. Therefore a ton of uranium-235 decays much faster than a ton of uranium-238. The speed at which each radioactive element decays is measured by its *half-life*, the time it takes for half a given number of nuclei of the element to decay. It makes no difference whether the quantity you start with is large or small; by the time one half-life is over, half of it will have decayed. The half-lives of the three radioactive elements most important in heating the earth are, for uranium-238, about 5 billion years; for uranium-235, 0.7 billion years; and for thorium-232, 14 billion years.

It follows that in the distant past all three elements were more abundant than they are now, and also that they must have been present in different proportions. For example, when the earth first came into existence, about 4.5 billion years ago, it must have contained slightly more than twice as much uranium-238 as it contains today, and the strongly radioactive uranium-235 must have formed a much larger proportion of the total radioactive material.

It is believed that 2.5 billion years ago and earlier, in the period called Archean time, a number of strongly radioactive elements were present that have since decayed to the vanishing point: they have gone extinct. The amount of radioactive heat generated then was probably three or four times as great as that generated now.[3] Regarding radioactive elements as a mixture of fuels, we can say that the total supply has dwindled considerably since the earth's early years and that some of the fuels are now wholly exhausted.

*The Earth's Original Heat, or Gravity Transformed*

Not all the earth's heat comes from radioactivity. Nearly as important (perhaps equally important) is the heat liberated when moving bodies—bodies imbued with kinetic energy—are brought to a stop. This happened in spectacular fashion when the earth came into existence. A cloud of rock fragments, dust, and gas was drawn together by the pull of gravity and accreted into a solid protoplanet;[4] as our protoplanet continued to grow, smaller, incipient protoplanets—large asteroids—that happened to be near, fell in on it and became part of it. All together, almost $6 \times 10^{24}$ metric tons of material crashed into one big lump—the earth. The force of gravity caused the multitude of contributing chunks to accelerate to high velocities as they fell, acquiring high kinetic energies in the process. When these energetic bodies collided, nearly all their combined kinetic energy (not absolutely all, as we shall see below) was converted to thermal energy—heat. The total energy was about $2 \times 10^{32}$ J; not surprisingly, the young earth melted.

When this happened, gravity again intervened, sorting the mixture of liquids by weight. The denser, heavier liquids, predominantly molten iron, gravitated down to the center, leaving the lighter silicates to float to the top and form the earth's outer layers. The tidy arrangement of fluid material, with the heaviest at the center and the lighter farther out, is what caused the earth to change from an irregular lump to a sphere. The sinking of heavier liquids and the rising of lighter ones also released still more gravitational energy, hence more heat.

Since Archean time, the earth has probably cooled several hundred degrees.[5] As the core cooled, the center, where the pressure was greatest, solidified first; in the outer layers, where the pressure was lower, the iron remained liquid. The outer core is still liquid, and as the planet keeps cooling the liquid continues to solidify, adding new layers to the inner core while the outer core thins correspondingly. In the act of solidifying, liquid iron loses some energy by liberating its latent heat of freezing (see chapter 10). This compensates for some of the heat being lost and slows the cooling.

At the surface of the core, where the temperature drops sharply, iron continually crystallizes; the solid crystals then sink through the outer core and come to rest when they reach the inner core.[6] This is still going on today, and it entails a continuing slow conversion of gravitational energy to heat; it is all part of an ongoing sorting process by which the earth's original ingredients, since they first accreted, have kept on shifting in a way that brings the densest material closest to the center.

In brief, the earth's primordial heat, even while it is constantly radiating away into space, is still being generated at a modest rate. We still have some residue of the heat of creation. But inevitably, in time it will all be gone.

*Spin Energy*

As we've noted already, the earth is believed to have formed from the gravitational collapse of a cloud of rock fragments, dust, and gas. The cloud was part of a much larger spinning, disk-shaped cloud—the solar nebula—that was the precursor of the whole solar system. The sun, the planets, and all the other bodies of the system, as they condensed from the parent nebula, inherited its spin; they all rotate, each at its own speed. In the case of the earth, the speed of rotation is once every twenty-four hours, as we all experience day after day.

Because it rotates, the earth has spin energy. The amount, close to $2 \times 10^{29}$J (see chapter 8), is a mere one-thousandth of the thermal energy liberated when the earth came into existence. All the same, it is part of the total energy that was bestowed on this planet at birth.

All the earth's energy is in the process of being dissipated; not to mince words, our home planet is steadily getting rid of its energy, shedding it, radiating it away. The way the 99.9 percent consisting of thermal energy is being shed is discussed in chapter 15. One of the ways the remaining 0.1 percent, the spin energy, is being shed is by tidal drag, as described in chapter 8. Drag is also going on deep inside the earth.

The flowing liquid iron of the outer core gives the earth yet another form of internal energy, magnetic energy (see chapter 16). This energy too depends on movement powered by heat; it will steadily diminish as the heat is radiated away and the flowing iron slows and hardens.

# 15 HOW THE EARTH SHEDS ITS WARMTH

## The Earth's Interior Is Cooling

The earth is constantly losing its thermal energy by all three of the mechanisms that transport heat from one place to another: conduction, convection, and radiation. Heat is conducted through the solid material—the inner core and the lithosphere (including the crust). It is carried upward in convection currents in the outer core, the mantle, the oceans, and the atmosphere, and it is radiated away into space.

Convection not only transports heat, it brings about the generation of more heat because of the inevitable drag within moving fluids. The resulting feedback makes convection the most complicated of the mechanisms of heat transport. Not surprisingly, the deepest of the convection currents, those in the outer core, are the least understood. The heat causing them comes, in unknown proportions, from several sources: radioactivity, gravitational energy, and the latent heat of freezing released by liquid iron as it solidifies. Heat from the core almost certainly assists the flow of convection currents in the mantle,

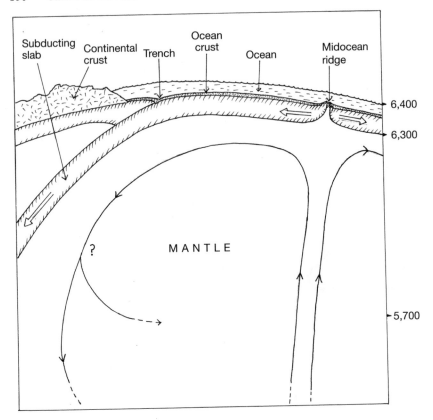

Figure 15.1. Tectonic plates separating (on the right), and one subducting below another (on the left). The open arrows show the direction of plate movement. The curved arrows represent the flow of currents in the mantle; the descending current may divide, as shown. Approximate distances, in kilometers, from the earth's center are shown on the right of the figure, which is not to scale. The same figure with added detail appears in figure 15.2.

which appear to be aligned with the currents in the core. The connection between core currents and mantle currents is at present unclear, however, because data from the deep inner mantle and the core are hard to obtain and difficult to interpret. Geophysicists and seismologists at work on the problem are testing a number of models to help them discover exactly what is happening thousands of kilometers down in the depths of the earth.

More is known about the convection currents in the outer mantle, which are coupled to the movement of the lithosphere's tectonic plates. The discovery of continental drift, as it used to be called, was one of the great advances of the twentieth century. With it came the realization that the pattern of the world's continents and oceans, far from being fixed for all time, is constantly changing; our mental image of the world was altered profoundly.

The way plates move is familiar nowadays from countless books and articles. Bear in mind that our focus here is on the energy associated with the movement. Figure 15.1 diagrams what happens. The curved arrows show the near-surface parts of the convection currents that bring about an overturn of the whole mantle, allowing heat to be transported outward, toward the earth's surface; the lower halves of the convection currents are left to the imagination, because their routes are enigmatic.

It is uncertain what keeps the convection currents going. Possibly they start as long, narrow plumes of hot material rising from the depths that, on reaching the base of the lithosphere, spread out and flow horizontally.[1] Possibly they start as upwellings sucked up to fill the gaps opening up as the tectonic plates separate. It is unclear whether currents in the mantle drag the floating plates or whether the drift of the plates keeps the mantle currents flowing.[2] In any case, the convection currents flow. A complete overturn of the mantle is believed to take hundreds of millions of years.[3] The power needed to drive plate motion is probably on the order of 1 trillion watts, or 1 terawatt.

The plates are believed to have been moving ever since the lithosphere first hardened. In Archean time, more than 2.5 billion years ago, when the mantle was hundreds of degrees hotter than it is now and its outermost layer (the asthenosphere) correspondingly less viscous, the plates must have moved faster, perhaps ten times faster, than at present.[4] The speed of the motion differs from plate to plate; the maximum speed at present is thought to be between 15 and 20 cm per year.

Hot, soft rock oozes up from the asthenosphere to fill the spreading gap wherever plates pull apart, notably at the midocean ridges. The upwelling molten rock (magma) cools and congeals and becomes part of the trailing edges of the retreating plates on both sides of the gap. It becomes new seafloor; the process is known as seafloor spreading. Because of the way new seafloor is continuously created, the seafloor near each midocean ridge on each side of it is younger and hotter than that farther away. The oldest and coolest part of a drifting plate is at its leading edge. The cool rock is dense, ready to sink into the mantle.

When two moving plates meet, one is driven downward and *subducts* below the other. If this happens in the ocean, a deep trench forms in the ocean floor where the edge of the subducting plate slopes steeply downward. The subducting slab travels with the descending part of the convection current. Studies of the speed of seismic waves traveling through the earth by different routes show that subducting slabs do not all follow the same path: some level off quite soon, about 700 km down, where the mantle rock becomes more viscous; others sink to much greater depths, down to the core-mantle boundary. A slab has been detected in the mantle below Siberia, at a depth of 2,800 km, that is estimated to be 200 million years old; it must have been in existence since very early in the time of the dinosaurs.[5]

It appears that the tectonic plates are pushed away from the midocean ridges where molten rock wells up; at the same time they are pulled toward the ocean trenches by the sinking of the cool, dense rock of the plates' leading edges—the subducting slabs. Two forces seem to be shifting the plates, which raises the question, Which is the stronger force? Are the plates pushed or pulled?

Research strongly suggests that the chief force is the pull of the subducting slabs.[6] It follows that "plate tectonics is a primary result of a cooling earth,"[7] because it is cooling that makes the slabs dense enough to sink, exerting a strong pull as they do so. The temperature difference between the newly formed, young crust close to the active parts of the midocean ridges and old crust about to subduct is sometimes 1,000°C.[8] A huge quantity of heat is evidently dissipated by the plates as they drift across the earth. It is heat they received from the mantle below them.

The surface of the earth is about 29 percent land and 71 percent ocean. Therefore most of the heat is carried away via the oceanic crust; that is, it is conducted through the crust, then convected to the ocean surface and finally radiated to space. The loss is most rapid on the seafloor where very hot, newly formed crust comes into direct contact with cold seawater. This happens along segments of the midocean ridges where the tectonic plates are separating most actively. Over time—geological time—different segments of a ridge become active; from time to time the activity dies down at one place and starts up at another. Submarine volcanoes erupt wherever the activity happens to be most vigorous; lava is extruded through volcanic conduits from kilometers deep in the crust, and cold water floods in; it is quickly heated by the hot rock to temperatures as high as 300 or 400°C (the high pressure at depth raises the boiling point of water even higher than this). The heated water escapes upward through separate vents, pipes, and cracks. In this way local *hydrothermal circulation systems* are set up within areas of the crust.[9] Huge volumes of water are circulated, so much that it takes only about 10 million years—a short period in geological terms—for the whole world ocean to pass through one or another hydrothermal system; this means that the whole ocean has circulated through hot new crust hundreds of times during the earth's lifetime so far.[10] Indeed, "the hydrothermal circulation is the cooling radiator . . . of the earth's engine."[11]

An individual hydrothermal system has a limited lifetime; as the seafloor spreads, it is carried away from the midocean ridge where it originated into a cooler region where it dies away. At the same time, a new hydrothermal system establishes itself in a new patch of hot seafloor, close to the ridge.

At the heart of an active hydrothermal system, the hottest water emerges through submarine geysers known as *hydrothermal vents,* which spew out superheated water with the force of a fire hose. The hot water rises into the surrounding cold seawater in opaque, billowing plumes known as *black smokers,* which look like the smoke from an oil fire. The color comes from suspended chemicals, mainly sulfides. Cooler vents creates *white smokers.* Warm water, cool enough not to scald you if you could put a hand in it, seeps out more gently through innumerable fissures.

The seafloor surrounding hydrothermal vents is the home of the *hydrothermal vent fauna,* a group of invertebrates that live in the earth's most recently discovered (1977) natural ecosystems.[12] There are no plants. The animals, many of them belonging to species new to science, live and thrive near the vents, at a safe distance from the scalding water but close enough to benefit from the warmth. The rich supply of sulfides in the water is the energy source for big populations of chemosynthetic sulfur bacteria (see chapter 12). The bacteria are at the bottom of the ecosystem's food chains; at the top are weird creatures such as giant tube worms, giant bivalves (clams and mussels), sea anemones, and eyeless shrimps. Spectacular photos and videos, taken from manned submersibles, have made the hydrothermal vent fauna familiar to a wide public.

Photos have also been taken, without artificial light, using an exceptionally sensitive camera. The photos show the underwater darkness to be relieved by a very faint glow, too dim to be perceived by humans but presumably bright enough to register with the seemingly blind shrimps, which have now been found to have light-sensitive organs on their backs.[13] How the energy for this faint light is generated is at present unknown; its source may be the chemical and physical reactions happening where superheated, mineral-rich water under high pressure emerges from the vents and makes contact with the cold water of the surrounding deep ocean.

Hydrothermal vents are a topic where the interests of geophysicists, oceanographers, biologists, and physical chemists converge. They are places where the earth's internal heat is being vigorously dissipated, where unique ecosystems flourish with no help at all from solar energy, and where light is produced in unusual ways.

## Turmoil at the Surface: Earthquakes

The energy trapped in the earth's interior, underground and out of sight, attracts no public attention until it bursts out of bounds. People who never go near volcanoes, or into earthquake zones, can live a lifetime without becoming aware of the vast stores of energy trapped below the ground; it has to be experienced to be appreciated. An earthquake is one of the ways the earth rids itself of a portion of the energy it must dissipate; other ways are volcanic eruptions and mountain building.

Let's consider earthquakes first. They occur when the edges on either side of a break in the lithosphere scrape past each other. The break may be along the contact between adjacent tectonic plates; this is true of most large earthquakes and explains why they happen at plate boundaries. Or it may be a fracture—a *fault*—anywhere within a plate where the rock has been deformed by pressure from a distant plate collision until it snapped.

Then, whatever the origin of the break, the masses of rock on the two sides of it slide past each other. The sliding isn't smooth. At first friction holds everything in place while stress builds up; eventually the force pushing the masses overcomes the friction, and they suddenly jerk past each other. The jerking movement—often a sequence of several jerks—is an earthquake shock and its aftershocks. The relative movement of the two masses may be up and down, sideways, or a mixture of the two. A vertical movement at the surface leaves a cliff—a fault scarp—as evidence; a lateral, sideswiping motion, such as happens along California's San Andreas fault, leaves unmistakable discontinuities in roads, fences, and streams.

In terms of energy changes, what happens is this: while they are held fast by friction, the rock masses accumulate *elastic* potential energy in the internal deformations caused by the tremendous pressure. Then, when the stress becomes too great, the frictional force locking them together fails and the masses slip—the potential energy becomes kinetic energy. The abrupt movement of massive bodies of rock is a *seismic event*—an earthquake. What becomes of the energy released?

It is dissipated in several ways: some is converted to  gravitational PE, stored in rock masses that are lifted to a higher elevation than they occupied before the quake. A fraction of the energy is used in smashing things—rocks and all manner of human constructions—which entails breaking chemical bonds. Another fraction is converted to thermal energy by friction at the place of the rock displacement. And what remains is transported away from the scene as *seismic waves*, to be dissipated, eventually, some distance from the site of their origin (see chapter 17).

Some earthquake energy, surprisingly, is added to the earth's spin energy.[14] The displacement of rock accompanying an earthquake usually shifts heavy rock downward, closer to the earth's center and thus closer to its axis of rotation than it was before the quake. In this way the earth's spin energy is being increased, at present, by about $21 \times 10^{13}$ kJ per year (this looks like a huge number, but it is only one-trillionth of the currently existing spin energy; see chapter 14). The increase causes the earth's spin to speed up, just as a spinning skater speeds up when she pulls in her outstretched arms. The faster spin would shorten the length of the day by an exceedingly small amount if the effect were not masked by the much more pronounced lengthening of the day caused by the drag of ocean tides.

The downward movement of a subducting tectonic plate under the pull of gravity also entails a loss of gravitational PE, in the same way that a landslide at the earth's surface causes surface rocks to lose gravitational PE (see chapter 9).[15] The principle is the same even though the details are spectacularly different. At the surface, a mass of rock loses its gravitational energy by toppling off a precipice, hurtling through the air at high speed, and crashing at the bottom. Inside the earth, a falling mass of solid rock sinks through ductile rock at a speed of only a few millimeters a year; as it sinks, it slowly warms to the temperature of the rock surrounding it until it becomes indistinguishable from it. In both cases—landslide and subducting slab—a small fraction of the lost PE becomes added to the earth's spin energy. And in both cases most of the gravitational PE is converted to KE, which in turn is converted to thermal energy because of friction and drag. In a word, it is dissipated.

*Turmoil at the Surface: Volcanoes*

Next, consider volcanoes. Obviously the earth loses heat when volumes of red-hot lava gush out through the crust and cool off in the open air. Not so obvious is the reason the lava became molten in the first place. Volcanic lava (known as *magma* before it emerges into the open) usually comes from no

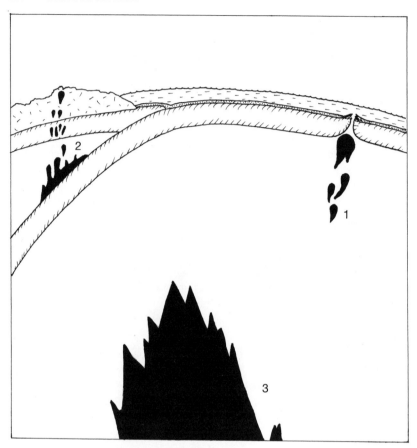

Figure 15.2. Ascending magma chambers full of magma (black) in the section shown in figure 15.1. Different kinds of magma chamber, labeled 1, 2, and 3, are formed by (1) the melting of ascending currents of mantle material; (2) frictional melting on the upper surface of a subducting slab; (3) the heat of a hot spot at the core-mantle boundary (far below the border of the figure).

great depth: it is molten mantle rock. It does not come from the core; although the outer core is liquid iron capable of flowing, it is far too dense to rise to the earth's surface. The problem becomes, Why should mantle rock melt?

Most volcanoes are near the boundaries of tectonic plates. Mantle rock melts, and the magma collects to fill *magma chambers*, in two very different environments: where two plates separate, and where two plates meet with one subducting below the other. The rock melts in these two environments for two quite different reasons. Figure 15.2 shows what happens (note the sites labeled 1 and 2).

Where plates separate, a current of ductile mantle rock ascends from below; as the rock creeps up, the pressure weighing it down decreases until the combination of pressure and temperature is such that the rock liquefies; this is the process feeding volcanoes on midocean ridges.

Where two plates meet, the subducting slab sinks down into the mantle on which it had been floating. Frictional drag between the cool, dense slab and the warmer upper mantle rock through which it sinks generates sufficient heat to melt the overlying warm rock but not the cool, sinking slab.[16] The reason the upper rock melts and the lower rock doesn't is that the rock above contains considerable moisture, absorbed from wet ocean sediment, and wet rock melts at a lower temperature than dry rock does.

Unlike most volcanoes, which are found at plate boundaries, some erupt nowhere near the boundaries. These are *hot-spot* volcanoes, fed by plumes of hot rock rising from much greater depths (site 3 in fig. 15.2).[17] The plumes originate at localized hot spots at the bottom of the mantle where it is heated to tremendously high temperatures by currents in the liquid outer core. The temperature difference between a plume and the rock around it may be as great as 1,500°C.

A hot-spot volcano is unaffected by the drift of the tectonic plates across the earth's surface nearly 3,000 km above the hot spot itself. In this respect hot-spot volcanoes behave quite differently than plate-boundary volcanoes do. The latter erupt wherever a boundary happens to be, because the magma sources travel with the plates; a hot-spot volcano, in contrast, is left behind by the plate moving over it; each time it erupts it punches a new hole through the ever-drifting crust and builds a new volcano some distance behind the site of the preceding eruption. As a result, the successive eruptions from a hot-spot plume create a row of volcanoes. The youngest volcano of the row is at the back of the line, with progressively older volcanoes ahead of it.

Rows of hot-spot volcanoes are seldom straight lines; more often they are gently curved arcs. Sometimes the arcs trail across a continent, sometimes across an ocean. If the summits of an arc of submarine hot-spot volcanoes are high enough to emerge above sea level, the result is a volcanic *island chain*. Several of them occur in the Pacific—for example, the Hawaiian Islands, the Aleutian Islands, the Kuril Islands, and the Tuamoto Archipelago.

Note that each of the three kinds of volcano we've considered receives its lava (or magma, while it's underground) from a different source. Midocean-ridge volcanoes get it from rising currents in the mantle, which melt when the pressure is low enough; the volcanoes above subducting slabs get it from moist, upper mantle rock that has been heated to its comparatively low melting point by friction; and hot-spot volcanoes get it from hot plumes rising from the bottom of the mantle. Note particularly that in no case does the melting require "new" thermal energy or energy from an outside source. The energy that volcanoes let loose has been inside the earth all along, waiting to be dissipated. There is plenty more down there still.

Generalizations about the energy released in volcanic eruptions would be meaningless because they are so variable. But it can safely be said that some past eruptions were larger by far than any experienced by human beings since our species evolved 3 or 4 million years ago. Eruptions two or three thousand times as powerful as the Mount St. Helens eruption of 1980 have left their mark on the earth.

On several occasions in the past, a newly developed hot spot has sometimes fed a long sequence of eruptions lasting for a few million years and leaving behind overlapping layers of volcanic rock (basalt) that now cover whole landscapes to considerable depths.[18] Each eruption obliterated every living thing in its path. Examples of the present-day remains of such events are the sheets of lava forming the Columbia Plateau of Washington and Oregon, which flowed out of the ground in a series of eruptions about 17 million years ago and buried more than 100,000 km² of land. A similar series of eruptions 65 million years ago produced the enormous terraced sheets of lava known as the Deccan that occupy most of peninsular India; their area is more than 500,000 km².

These sequences of tremendous lava floods have happened at long intervals, on the order of tens of millions of years. Another such sequence could begin anywhere at any time. The earth still has plenty of energy to dissipate.

*Mountain Building*

Volcanic eruptions and earthquakes are local events in which the earth's energy is dissipated in short, spectacular bursts. At the same time, the earth is continuously shedding huge amounts of energy in the long, slow process of mountain building. It happens when the continents borne by tectonic plates collide and deform each other. The upper layers of rock are compressed, and the way they respond depends on their structure.

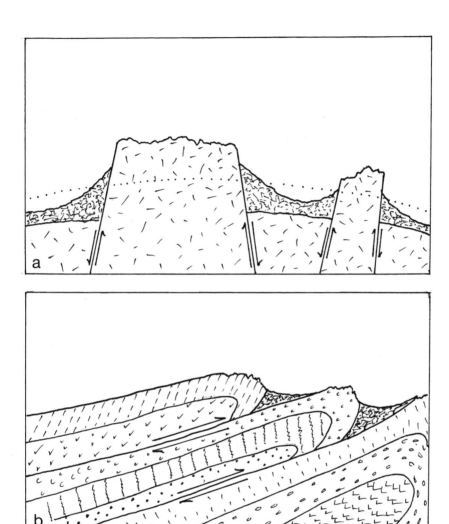

Figure 15.3. (a) Mountains built by normal faulting; the dotted line shows the surface before the faulting. (b) Mountains built by the folding and thrust faulting of sedimentary strata. In both diagrams, half arrows show the direction of relative movement at each fault; recently eroded material in the valleys is shown scribbled.

Two possible outcomes are beautifully illustrated in the Rocky Mountains, which were forced up, and continue to be forced up, by the push of the Pacific plate against the western margin of the North American plate.[19] In the southern Rockies (Colorado and southward) the pressure has forced the crust to arch upward. Tension over the top of the arch has caused vertical or near-vertical cracks (*normal faults*) to open up, separating the crust into a number of steep-sided blocks, some of which have been driven farther upward by compression; many of these blocks, consisting of ancient granite from which overlying sediments have mostly been eroded, now form precipitous mountain peaks (fig. 15.3a).

In the northern Rockies, by contrast, the compression forced thick, horizontal layers of sediments to slide forward over the underlying rock. Friction resisted the slide and forced the sheets of rock to buckle into a series of folds. Under continued pressure, many of the folded sheets fractured and were thrust forward and upward over the sheets ahead of them along gently sloping *thrust faults;* the overlapping thrust sheets stacked up to form mountains several thousand meters high (fig. 15.3b). This is how the folded mountains of Wyoming, Montana, and Alberta came into existence.[20]

Both processes obviously entail the dissipation of enormous amounts of energy. The rock faces on either side of a fault grind past each other against unimaginably strong frictional resistance. Most of the movement probably happens in sudden spurts, separated by long intervals; frictional heat is generated whenever there is movement.[21] The folding of rock consumes much energy too. It takes place deep underground, where the high temperature makes the rock plastic enough to bend rather than break under intense, slow-acting pressure. Energy is consumed in stretching and rupturing the intermolecular bonds that hold the rock together, leaving it permanently deformed.

As rock is raised to a higher elevation in the process of mountain building, it automatically gains gravitational potential energy (see fig. 15.4); in this way some of the KE of drifting tectonic plates becomes converted to PE and "stored" in the mountains as they rise. It doesn't all remain stored, though. Some of it is dissipated as it accumulates, because a growing mountain sinks to some extent at the same time as it grows. In the act of sinking, the crust dissipates the surplus energy (gravitational PE) in two ways. In the first place, the added weight of the growing mountain causes the lithosphere supporting it to sag; the lithosphere is an elastic solid, able to bend without breaking under pressure that does not exceed its *elastic limit.* The sagging lithosphere stores

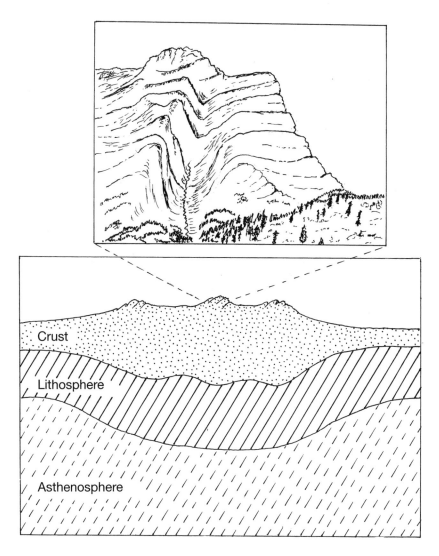

Figure 15.4. Lower panel: The weight of mountains at the surface stretches the elastic lithosphere supporting their weight. The lithosphere sags into the viscous asthenosphere it floats on. Upper panel: Sketch of one of the folded mountains in the diagram below (Mount Kidd, Alberta, 2,972 m high).

energy as elastic PE; then, as the load is removed by erosion, the elastic PE energizes an exceedingly slow "rebound."

In the second place, the sag in the floating lithosphere displaces some of the athenosphere it floats on (see fig. 15.4). In the same way that an iceberg floating in the sea displaces its own weight of water, the sagging lithosphere displaces its own weight of viscous asthenosphere material: it gives the material the energy to ooze to one side.[22]

We have now completed the energy budget for mountain building. To summarize: the source is the kinetic energy of drifting tectonic plates. It is dissipated in several ways: as frictional heat, when faults slip; in the breaking of chemical bonds, when rocks are forced into permanent folds;[23] and as gravitational potential energy temporarily parked in the uplifted masses of rock. This gravitational PE is quickly (in geological terms) transformed, some to elastic PE in the sagging lithosphere and some to the KE of the oozing asthenosphere material displaced by the sagging lithosphere. Ultimately, erosion removes the mountain, the lithosphere rebounds, releasing its stored elastic PE, and the asthenosphere oozes back to where it was before; drag causes a large fraction of the asthenosphere's KE to be dissipated as heat.

The energy that went into building the mountain is now all accounted for. The most difficult part—putting in the numbers—will not be attempted. Possibly the biggest difficulty is to assess how the PE stored for a while in the uplifted mass of the mountain becomes apportioned between elastic PE in the lithosphere and KE in the asthenosphere.

# 16 ELECTROMAGNETIC ENERGY

*Action at a Distance*

In this chapter we look into a form of energy so far mentioned only in passing. To begin, it is well to repeat, in different words and with less detail, three statements from chapter 2.

First, applying a force to an object—pushing it or pulling it—changes the object's motion; it either accelerates it or decelerates it in the direction of the force. If the object was originally stationary, the force causes it to start moving—accelerates it.

Second, gravity is a force. Specifically, it is the force that causes any mass to attract toward itself any other mass at a distance from it. The example known to virtually everybody is the traditional tale of Newton and the apple; gravity between the gigantically massive earth and a comparatively tiny apple (growing on Newton's apple tree) caused the apple, which wasn't firmly attached to the tree, to fall to earth. Simultaneously, gravity also caused the earth to move an imperceptibly small distance toward the apple, but this point is rarely mentioned because the effect is far too minute to be measurable. Note that gravity acts at a distance—the masses that attract each other may be widely separated: more on this below.

Third, a force acting through a distance performs work (in the physicist's sense of the word) or, equivalently, expends energy (again in the physicist's sense). This is what defines both *work* and *energy*.

These statements immediately invite two questions: Do any other forces resemble gravity in acting at a distance? And exactly how does action at a distance operate?

To answer the first question, Yes, there are other familiar forces that act at a distance, namely, electric force and magnetic force. At one time they were thought to be separate and unrelated, but nineteenth-century physics showed that in fact the two seemingly different forces are the outcome of a single physical process. Everybody is familiar with them even if they haven't thought much about them. Anyone who has experienced static cling has seen electric force in action, and anyone who has seen a refrigerator magnet in use has observed magnetic force in action.

We go further into these matters after tackling the second question: How does action at a distance work?

Take gravity as a specific example. Its action can be explained in various ways. One explanation is that any piece of matter—any mass—is surrounded by a *field of gravitational force* or, more briefly, a *gravitational field*. This means that at any chosen point in the space surrounding the given mass a force of specifiable strength and direction will act upon some other mass, used as a "test" mass, if you place the test mass at the chosen point (see fig. 16.1). This happens because an unseen gravitational force pervades the space surrounding the given mass in the form of a *force field* reaching out, in theory, to infinite distance. If this is an acceptable "explanation" of gravity, it leads to the conclusion that the given mass doesn't really act at a distance. Rather, it is the force field that affects the test mass, by acting on it directly at the precise spot where field and test mass touch. Inspecting this explanation shows that in fact it isn't one. It merely replaces the notion of a force acting at a distance with a force field acting wherever you want, leaving the latter concept still undefined.

One model of the way a force field acts follows from Einstein's general theory of relativity: it is that forces act "downhill" within a "space-time" having a "geometry" modified by the presence of masses scattered here and there within it. This is all very well, but the only outcome of the foregoing arguments are (fairly) easily visualizable mental images of how gravity acts. More modern images, designed to explain gravity in the context of modern quan-

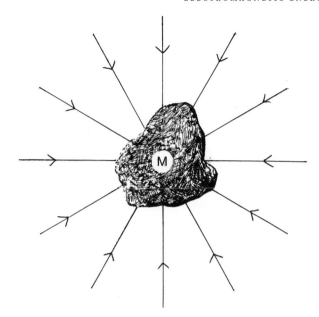

Figure 16.1. The arrows show a few representative lines of force of the gravitational field surrounding a chunk of rock, M, isolated in space.

tum theory, are more sophisticated than their predecessors, but they are still only mental images, as physicists readily concede.[1] It seems futile to keep refining them before one has arrived at a satisfactory definition of what is meant by a body's *mass*, a term we have not yet defined. Here is a modern definition: the mass of a body is a measure of its resistance to being accelerated or, what comes to the same thing, its resistance to a force.[2] The circularity is seamless: a force is something that acts on a mass, while a mass is something that responds to a force.

It therefore seems best to treat both "mass" and "force" as terms labeling fundamental, intrinsically undefinable concepts that must be accepted a priori in order to make further discussion of the material world possible: one has to start somewhere. In what follows, we let the two terms have their obvious, intuitive meanings.

To return to the first question we asked above, Do any other forces resemble gravity in acting at a distance? As already remarked, two other forces behave like this—electric force and magnetic force. To begin with we consider them separately, deferring for now the linkage between them.

*Electric Force*

A weak electric force is easy to generate. Most children have seen it done as a party game. An effective method is to rub a plastic rod (a ballpoint pen or the handle of a plastic spoon) on a piece of fur (human hair serves well); rub vigorously for a minute or two and then promptly hold the end of the rod just above a few small scraps of torn paper. One of the scraps will rise and stick to the tip of the rod, and other scraps will follow until several are lined up as if strung together, end to end. Evidently a force is at work strong enough to overcome the force of gravity that held the scraps down on the table; and the force acts at a distance. Admittedly the force is slight (the scraps of paper aren't heavy), and the distance over which it acts is a centimeter or two at most; all the same, a force acting at a distance has been created.

What has happened is that the brisk rubbing has dislodged some outer electrons from atoms at the surface of the fur and left them adhering (temporarily and weakly) to atoms at the surface of the plastic. The adhesion is a frail version of an ionic chemical bond as described in chapter 10. In acquiring extra electrons, the plastic has acquired a negative electric *charge;* simultaneously, in losing these electrons, the fur has acquired a positive electric charge.

The plastic is now capable of exerting an electric force that attracts positively charged objects and repels negatively charged ones. But the scraps of paper have not had electrons rubbed off them or stuck to them—they are electrically neutral: Why should they be attracted to the negatively charged plastic rod? The answer (see fig. 16.2) is that the negative charge on the tip of the rod repels electrons from the near edge of the closest scrap, giving this edge a positive charge so that it is attracted to the rod. The farther edge of the same scrap receives the repelled electrons, giving it a negative charge: thereupon it acts on the second scrap of paper in the same way that the plastic rod acted on the first scrap. This chain reaction can seldom be made to reach beyond three or four scraps of paper because stray electrons "leak" between the charged paper and the surrounding air: the small electric charges soon fade away.

This desktop experiment allows you to produce and examine one of the fundamental forces of nature, the electric force that holds atoms and molecules together (but not atomic nuclei). The force "acts at a distance" as gravity does, but it is wholly unlike gravity in acting as both an attractive force and a repulsive force. Gravity has an identical effect on *all* masses: it is always a force of attraction; it never causes one mass to repel another. In contrast, electric force acts only on electrically charged bodies, and it acts in two ways: as a force

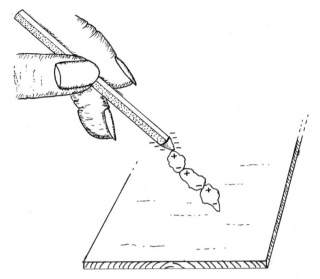

Figure 16.2. A rubbed plastic rod attracting a series of paper scraps. Positive and negative charges are shown by the + and - signs.

of attraction between two bodies having unlike charges (one positive and one negative) or as a force of repulsion between two bodies having like charges (both positive or both negative). Note also that gravity acts on all material bodies, for all of them have mass, whereas electric force acts only on electrically charged bodies; it has no effect on uncharged, electrically neutral ones. An electrically charged object has either an excess of electrons, giving it a negative charge, or a shortage of electrons, leaving it positively charged.

Now consider what happens to the surplus electrons on a negatively charged object. Each one repels all the others. Can they "escape"? The answer depends on whether the object is connected to the ground, however indirectly, by materials that conduct electricity (*conductors*), or that do not (*insulators*).[3]

If the object is separated from the ground by good insulators, then it will retain its charge—the surplus electrons—for quite a long time (not forever: no insulator is perfect). Because the electrons repel each other, they will come to be evenly spread over the surface of the object and will remain there; in a word, they will be *static*. This is why the electric charges, forces, and fields we have been considering are often called *electrostatic* charges, forces, and fields. While it is charged, the object will be surrounded by an electric field in the same way that an object with mass—any object, in fact—is surrounded by a gravitational field.

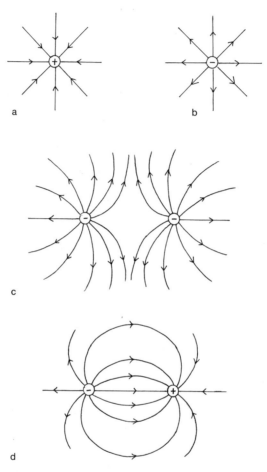

Figure 16.3. Electrostatic force fields surrounding (a and b) a single charge; (c and d) a pair of charges. The charges of the bodies are shown. In each case the arrows point in the direction a negatively charged test charge would move.

Figure 16.3 shows some typical electric fields. The lines are *lines of force.* Each line is the direction (shown by the arrows) that a small, negatively charged body—a test charge—would move if placed on the line. In theory, infinitely many lines could be drawn; in practice, only enough are shown to illustrate the form of the field without cluttering the drawing.

Imagine next that the negatively charged object is connected to the ground by a conductor—say a metal rod. The surplus electrons it carries will quickly flow to the ground through the metal, where they will be instantly absorbed and neutralized. All metals are good conductors: a characteristic of metals is that their electrons "are not held permanently in orbits related to particular atoms but can rove freely . . . . They form what is sometimes known as an `electron sea.'"[4] Consequently, electrons can travel quickly and easily through them, creating an electric *current*. Rapidly flowing electrons and an electric current are, indeed, the same thing. The passage of about $10^{18}$ electrons per second through a conductor is a current of one *ampere*, or amp, the unit in which current is measured.[5] Note that the number is only approximate: an ampere is not defined in this way, but rather in terms of the force the electrons exert. A precise definition appears in a later section.

We have now reached a point where it is possible to consider the energy provided by an electric current. The force that acts whenever an electric current flows generates energy, and the energy can be dissipated in a variety of ways. Often it is given off as heat: the filament in a lightbulb heats up when current flows through it, likewise the element in a toaster or an electric radiator (the way heat and light are radiated is discussed in chapter 18).

The rate at which an electric current yields energy—its power—doesn't depend only on the current (the number of amperes flowing). The force driving the current—the voltage—is equally important.[6] Imagine a waterfall and regard the water as analogous to an electric current flowing along a conductor. The power of the waterfall depends on both the volume of falling water and the distance it falls. With an electric current, the number of amps corresponds to the volume of water, while the voltage corresponds to the height through which it falls.

The voltage between two points on a current-carrying conductor—a wire, for instance—is said to be one volt if a current of one ampere yields power equal to one watt (one joule per second). This leads to the well-known formula

$$\text{amps} \times \text{volts} = \text{watts},$$

which can equally well be written

$$\text{volts} = \text{watts} \div \text{amps}.$$

The latter formula leads to the definition of a volt: one volt is the voltage between two points on a wire carrying a current of one ampere, when the power dissipated between the points is one watt. As promised, the definition of an ampere will appear later. One other unit can conveniently be defined here—

the *electronvolt* (symbol eV). It is the energy gained by a single electron in "falling" (recall the waterfall analogy) through one volt.

The concept of potential energy—energy ready to be dissipated when circumstances make it possible—is as relevant in the context of electricity as it is in the context of gravity, described in chapter 2. Think of the waterfall analogy again. We know that gravitational potential energy (PE) resides in a mass that would fall if it were able to, for example, a massive body of water prevented by a dam from flowing down a valley. Similarly, electric PE resides in an electrical charge prevented by insulators from going anywhere.

## Lightning

As illustrative examples of electric fields, we have so far considered only small-scale (very small scale!) fields that are artificially created indoors. Now we turn to the electric fields occurring outdoors in the natural world. They give unmistakable evidence of their presence every time lightning strikes.

In calm weather, the surface of the earth is negatively charged: a permanent electric field is believed to encase the whole earth.[7] The voltage across the gap between the *ionosphere* and the ground is estimated to be anywhere between 200,000 and 1,000,000 volts. The ionosphere is the electrified upper atmosphere starting about 100 km up. A current of one or two *picoamperes* per square meter is thought to flow continuously across the gap (one picoampere is $10^{-12}$ amperes); although air is a very good insulator, it is not so perfect as to prevent the flow of a current as small as this.

Lightning strikes when high voltages develop in the lower atmosphere, between a cloud base and the ground and between adjacent clouds (fig. 16.4). The way these strong electric fields develop is the subject of ongoing research; collecting data is both difficult and dangerous. All that needs to be said here is that it is hardly surprising that electric charges are apt to develop on the myriad tiny particles always suspended in the air. They are in a medium that is both turbulent, causing the particles to collide, and insulating, enabling them to retain the charges they acquire by losing or gaining electrons when they strike each other. Voltages high enough to produce lightning develop best in clouds with abundant ice particles.

Finely divided dust in a turbulent matrix is also found in the plumes of erupting volcanoes; the dust particles acquire electric charges, and lightning flashes are sometimes seen within the plumes.

In thunderstorms, lightning strikes when the voltage becomes so high that

Figure 16.4. The arrangement of electric charges, in a thunder cloud and on the ground and buildings below, when lightning is about to strike. The negative charge at the bottom of the cloud repels the surplus electrons on the surfaces directly below (ground, trees, buildings, and the like), leaving them positively charged.

the air fails to function as an insulator.[8] A large current flows, momentarily, through a fraction of a kilometer and produces a blinding flash. The current may flow between a cloud and the ground or between neighboring clouds; in either case the charge that created the voltage is discharged (neutralized), whereupon new charges quickly develop. Details on the exact behavior of lightning can be found in any book on meteorology, but it is worth remarking that the reported quantities—volts, amps, watts, and joules—are wildly inconsistent in the different accounts, because of the difficulty of making the appropriate measurements. A lightning flash comes unexpectedly, it is over in a fraction of a second, and its peak power would overwhelm ordinary measuring instruments; currents and voltages have to be inferred from indirect measurements. Often the quantities reported are not comparable: for instance, it is impossible to convert from joules to watts (joules per second) or vice versa without knowing how long a flash takes to complete. Currents are said to range from several thousand amperes to several hundred thousand.[9]

A quantity on which there seems to be some agreement is the temperature to which a lightning flash heats the air around it—about 20,000 to 30,000°C, which is three to four times the temperature at the surface of the sun.[10] The sudden heating causes an explosion of the heated air as it expands. This explosion and its reverberations produce a sharp clap of thunder and the rumbling that follows it; we consider the energy in sound waves in chapter 17.

*Magnetic Force*

Anybody who has refrigerator magnets uses magnetic force every day without thinking about it. To focus one's thoughts, it helps to do a simple desktop experiment like the one shown in figure 16.2, but this time employing magnetic, rather than electric, force. The instruments needed are a few sewing pins and a straight bar magnet (most refrigerator magnets are poorly shaped for the test and too strong; an adequate bar magnet can be made by repeatedly stroking an iron nail with a strong refrigerator magnet, always in the same direction). You can then pick up a series of two or three pins as shown in figure 16.5; note the strong resemblance to figure 16.2. It is evident that the tips of the nail and the pins acquire a property akin to electric charges, of two kinds.

Every magnet has two dissimilar *poles,* and as with positive and negative electric charges, unlike poles attract each other whereas like poles repel each other. This is easily tested using two straight magnets. The poles of a magnet are known as its north and south poles, and every magnet, without exception, has one of each. There's no need to know which is which to carry out the test if you use one end of a handheld magnet as a probe to "explore" a second magnet; you find that the probe attracts one end of the second magnet and repels the other end. The test can conveniently be done using a compass needle as the second magnet. The compass needle is itself a magnet mounted so that it can swivel freely; it  spontaneously aligns itself with the magnetic field of the earth. This is why a magnet's poles are labeled north and south, or N and S for short.[11] What makes the whole earth into a single huge magnet is a topic we come to later in this chapter.

We now consider the differences between the phenomena of figures 16.2

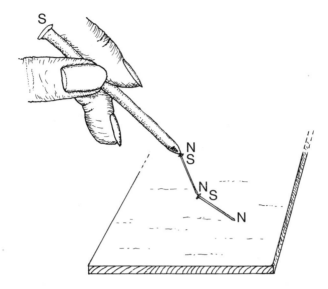

Figure 16.5. A magnetized nail picking up two pins. North and south poles are labeled N and S. (See text for details.)

and 16.5. The most obvious is that in the experiment with plastic and paper the materials are insulators, whereas in the experiment with nail and pins they are conductors. This shows that the same force cannot be responsible for both phenomena. Being good conductors, iron nails and pins cannot hold electrostatic charges and therefore cannot exert electrostatic forces.

Another, less obvious difference is that whereas electric charges can exist independently of each other, magnetic poles cannot. To demonstrate this requires more materials than the desktop experiments described so far; you may need to accept the following descriptions on faith. Suppose two plastic beads are rubbed with fur and two glass beads with silk (it is known, from separate tests, that the plastic beads become charged negatively, the glass ones positively).[12] Each bead is hung from a length of thread, and pairs of beads are brought close to each other. The two plastic beads repel each other; likewise the two glass beads. But if one of the glass beads is brought close to one of the plastic ones, they attract each other. This is more than simply a demonstration of the phenomenon already noted, that like charges repel each other and unlike charges attract each other. In this respect they behave exactly like magnetic poles.

The noteworthy difference between the two experiments is this: electrically charged bodies can exist separately and independently of each other; magnetic poles cannot. Thus any one of the charged beads in the experiments just described can be carried into another room and then brought back without its charge disappearing or being altered. In contrast, the poles of a magnet, *cannot* be separated from each other by any means whatever. All magnets have a pole of each kind, a north pole and a south pole. If you cut a bar magnet in half, new poles will appear at the cut ends, so that the  original magnet has become two half-length magnets, each with a pair of opposite poles, like this:

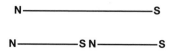

The final difference to note is that rubbing glass or plastic creates an electric charge, and with it an electric field, where none existed before. The energy of the rubbing is stored in the electric field. No analogous method will create a magnet: magnetic poles cannot be made to appear just by using muscle power.

Where, then, do magnets come from? How are they created? Weak magnets—lodestones—occur naturally and have been known for at least 2,500 years; a lodestone will attract other lodestones and also pieces of iron. A compass constructed from a sliver of lodestone was probably first used for overland navigation about 1,000 years ago, and for ocean navigation not long after. Lodestones consist of the mineral magnetite, an oxide of iron that is nowadays used as an iron ore in places where it is abundant.[13]

The problem of why some materials are strongly magnetic, or in technical terms, *ferromagnetic*, is one that would take us—if we were to follow it—deep into modern quantum theory. The elementary (and very incomplete) answer is that every electron is a miniature magnet.[14] Therefore any chunk of material, of any kind, contains countless hordes of subatomic "electron-magnets." In most materials these electron-magnets are all aligned independently, pointing in every possible direction, so that taken together they cancel each other out, leaving the material as a whole nonmagnetic.

In *ferromagnetic* materials the electron-magnets, instead of acting independently, line up with each other spontaneously in microscopically small "packets" known as *domains*.[15] This makes the material behave as an ordinary iron nail does in the presence of a ready-made magnet: although it is attracted by the magnet, the nail  will not act as a magnet itself unless it is in contact with a ready-made one.

Ferromagnetic material can be turned into a magnet proper by putting it into a magnetic field strong enough to line up all the little domains so that they become parallel with each other (hitherto they had been pointing randomly in all directions). Once they are aligned, they stay aligned: the material has been converted into a "permanent" magnet. This is what happens when you stroke an iron nail with a magnet as described above: the nail, though not a magnet, originally consisted of a vast number of unaligned magnetic domains. Stroking it with a strong magnet turns all the domains in the nail so that they are aligned along it, all pointing in the same direction. This means that the nail has itself become a magnet; it will continue to be one, if not permanently, then until it is melted, violently hammered, or otherwise mistreated.

## The Link between Electricity and Magnetism

It isn't necessary to have a magnet to create a magnetic field. it can easily be done by connecting the ends of a length of copper wire to the terminals of a six-volt electric battery (cautionary note: this short-circuiting of a battery should be done only momentarily, to avoid overheating). Before closing the circuit, lay the wire flat on a table and place a compass on top of it. Before the circuit is closed, the compass needle points to magnetic north in the usual way: it is unaffected by the nonmagnetic copper wire touching it. But as soon as the circuit is closed, the needle will align itself at right angles to the wire, leading to the conclusion that the flowing current is creating a magnetic field.

Figure 16.6 contrasts the ambient magnetic field due to the earth's magnetism far from electrical disturbances with the field near an electric current. In figure 16.6a a collection of compasses (only their needles are shown) is arrayed on a table; they all align themselves parallel with one another, pointing toward  magnetic north. Figure 16.6b shows the same setup, but this time a current-carrying wire passes through a hole in the middle of the table; the wire is at right angles to the tabletop, and its cross section is the black dot. Now the compass needles align themselves with the magnetic field around the live wire. As they clearly indicate, the lines of magnetic force form concentric circles centered on the wire.

Note that the lines of force are closed loops, with no end points. Contrast them with the lines of force of a gravitational field, which all terminate on a mass (fig. 16.1), and with the lines of force of an electric field, which all terminate on an electric charge (fig. 16.3). This suggests that there must be some connection between the impossibility of separating the poles of a magnet and the

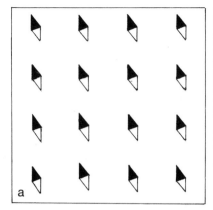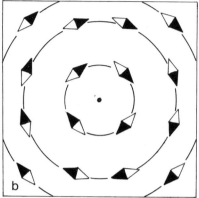

Figure 16.6. An array of compasses (a) in the earth's natural magnetic field and (b) in the field surrounding a current-carrying wire. The wire enters the page at right angles; its cross section is shown by the black dot.

fact that the magnetic lines of force around a current-carrying wire form closed loops; they do this invariably, however coiled or tangled the wire may be.

To explain why this should be so without going into details (which would take us too far from the subject of energy), we need only consider the field of force of an iron bar magnet, shown in two forms in figure 16.7. Figure 16.7a shows how the lines of force would be interpreted if there were indeed such a thing as a "magnetic charge" at each end of the magnet, where the lines of force appear to terminate. Figure 16.7b shows the lines of force as closed loops; from the point where each line seems to end at the magnet's south pole, it is assumed to continue, through the interior of the iron bar, and emerge at the north pole. The lines are believed to "thread through tiny circulating currents on an atomic scale."[16] Indeed, these are the currents that cause the magnet to be a magnet.

Because a current-carrying wire creates a magnetic field around itself, it functions as a magnet. Two current-carrying wires ranged side by side exert magnetic force on each other. If the current flows in the same direction in both wires, they attract each other; if the currents are antiparallel (oppositely directed), they repel each other.

Now for the promised definition of an *ampere*, the unit used for measuring an electric current: if the magnetic force between two identical straight, parallel, current-carrying wires separated by a gap of one meter is $2 \times 10^{-7}$

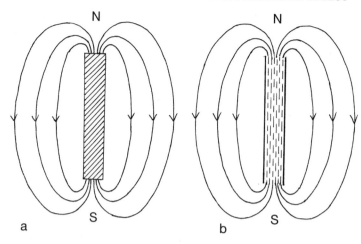

Figure 16.7. (a) The observed field of force of a bar magnet. (b) The (inferred) complete field of force. All lines of force are closed loops passing, for part of their length, through the solid iron of the magnet.

newtons per meter of length, then the current in each wire is defined as one ampere (recall from chapter 2 that one newton is the force required to give a mass of one kilogram an acceleration of one meter per second per second). An ampere is a measure of current or, equivalently, of moving electric charge. Next, we need a quantitative measure of electric charge itself. The unit devised for this is the *coulomb:* one coulomb is the amount of electric charge transported in one second by a current of one ampere. These units are named for two French scientists whose researches, in the eighteenth and nineteenth centuries, helped unravel the connection between electricity and magnetism.[17] That electric charge is measured in terms of electric current, which is measured in terms of magnetic force, gives some idea of the order in which different topics were developed.

The knowledge that flowing electric charge (a current) creates a magnetic field leads to the suspicion that a moving magnetic field might create a current. It does. The most convenient way to make it happen is to move a conductor (a wire or a coil of wire) through a stationary magnetic field; this causes an electric current to flow through the conductor. The kinetic energy provided to the conductor by whatever is moving it becomes converted to electrical energy. This is how an electrical generator (sometimes called a dynamo) works.

The close relation between electric and magnetic forces should now be clear.

A motionless electric charge (an electrostatic charge) creates an electrostatic field, while a stream of electric charges (an electric current) creates a magnetic field. The discovery that electric fields and magnetic fields are actually two manifestations of a single phenomenon, now known as an *electromagnetic field,* was one of the greatest scientific advances of the nineteenth century. We return to the topic in chapter 18. Before doing so, we look at the magnetic field of the whole earth, which enfolds all of us, everywhere and all the time.

## The Earth as a Magnet

A hiker using a compass is benefiting from the fact that the whole earth is a magnet. The compass needle, because it is a magnet, spontaneously aligns itself with the earth's magnetic field.

This raises the question, Why should the earth be a magnet? That it is one implies, as we saw in the preceding section, that electric currents must be flowing somewhere around or within the earth. Where and why? And what is the source of the energy driving the currents?

Theoretical arguments show that the currents must be within the earth. Moreover, they must be confined to the iron of the core, because the mantle consists of silicate minerals, which are electrical insulators.[18]

It is believed that convection currents in the liquid layer of the core rotate in a manner that generates a magnetic field that is almost (but not quite) parallel with the earth's axis of rotation. At the same time, electric currents flow in the liquid iron because it is moving in the magnetic field. Positive feedback is always in progress: the electric currents boost the magnetic field, and the magnetic field intensifies the electric currents.

The whole subject—*magnetohydrodynamics*—is made excruciatingly complicated by, among other things, the reciprocal interactions between the currents *of* liquid iron flowing convectively and the electric currents that flow *in* the iron. The words "current" and "flow" must both be used in two senses to describe what is going on. Suffice it to say that the energy generating all this action comes from the radioactivity that heats the earth's core, plus the residual heat still remaining from the time of the earth's formation. As we saw in chapter 15, heat from these sources is what keeps the outer core molten and convection happening; ultimately, it keeps the earth's magnetic field in existence.

# 17 WAVE ENERGY

## SOUND WAVES AND SEISMIC WAVES

*What Are Waves?*

Waves have cropped up in several contexts in this book. Chapter 7 described ocean waves—or more generally waves moving across a surface. Chapter 15 mentioned earthquake waves. Sound waves have appeared in a variety of contexts, for example, the roar of thunder and of landslides and the crash of breaking waves. Solar radiation, the chief source of energy at the earth's surface, has been touched on in contexts too numerous to list; likewise the infrared radiation emitted by sun-warmed land and sea. Electromagnetic radiation—light, radiant heat, and many other varieties—consists of *electromagnetic* waves; they will be considered in detail in chapter 18.

Waves of various kinds account for much of the energy of nature; indeed, they account for most of the energy we experience with our senses. All that we see comes to us as light waves and all that we hear as sound waves, and much of the warmth we feel comes as radiant heat. The output from radio and TV transmitters also comes into our houses uninvited, as imperceptible electromagnetic waves ready to be converted to sound and light waves if we choose to listen and look. Waves (strictly speaking *traveling waves*) demand examination in any discussion of energy.[1]

The first questions to be considered are, What are the different kinds of waves, how do they differ from each other, and what, in spite of the differences, unites them all as waves? What are waves and what do they do?

Here are two definitions of the word "wave" from physics dictionaries: "a periodic change in some property or physical quantity through a medium or space," and "a disturbance which propagates from one point in a medium [or empty space] to other points without giving the medium [or the space] as a whole any permanent displacement."[2] These definitions combined give us what we want. Thus water waves are displacements of a water surface to give regularly spaced crests and troughs that travel across the surface leaving the water unaltered and unmoved; in other words, leaving no trace. Similarly, sound waves are local pressure changes in a medium (usually air) that travel through the medium leaving no trace. Earthquake waves—*seismic waves*—often do leave a trace, but weak seismic waves—mild tremors—usually don't. Electromagnetic waves are very rapid changes in electromagnetic fields, moving at unimaginable speed through empty space and, once again, leaving no trace.

In all cases a physical quantity such as water level, air pressure, or the like varies regularly in a direction leading away from the source of the waves: the series of variations can be observed if you take an instant snapshot of it, in fact or in imagination. The same variations can be detected by a measuring device (or simply by eye if you're watching ocean swells) focused on a fixed spot for a length of time: the displacements or disturbances are then detected at regular intervals. To recap, for emphasis: a series of waves can be observed spread over space at one instant, or spread over time at one location. And waves are moving disturbances in a medium or in space: the medium itself (when there is one)does not shift in the direction the wave is traveling.

The preceding paragraph summarizes what waves are. Next, what do they do? They convey energy from one place to another. Equivalently, in waves "energy move[s] from one point to another but no material object makes that journey."[3] You can confirm this by tossing a rock into a calm pond and contemplating the circles of waves that spread out: the waves carry a portion of the kinetic energy the rock possessed at the moment it touched the water, and they carry that energy outward in all directions away from the point. The passing waves leave the water almost unchanged by their passage. The word "almost" allows for the fact that an immeasurably small fraction of the waves' energy is inevitably converted to heat because of viscous drag in the circling water under each wave (see fig. 7.3). The rest of the energy is carried to the margin of the pond (provided it's a small pond), where it sways water plants and shifts clods of mud. Notice that the energy the tossed rock once possessed has reappeared as movement at the margin of the pond: the waves have performed "action at a distance," without any horizontal displacement of the water.

The distances across which waves can transport energy are sometimes enormous. Earthquake waves can travel from their starting point to the other side of the earth (not necessarily in a straight line, as we shall see below). Giant sea waves—tsunamis—have been observed to travel 17,000 km (see chapter 7). Sound waves can travel 20,000 km in the sea, given the right conditions. Electromagnetic waves from quasars[4] reach the earth from billions of light years away, action at a very long distance indeed.

But no wave can travel forever; all are eventually dissipated, their energy degraded to entropy. In the context of energy, every kind of wave inspires two questions: How do they originate? And how are they dissipated?

## Sound Waves

Perhaps the most familiar waves, apart from water waves, are sound waves. Like all traveling waves, they are generated at one location and dissipated at another.

Take a simple example. An easy way to generate a sound—to make a noise—is to strike one object with another, for instance, to hit a nail with a hammer. Imagine a nail is being driven into a block of wood and focus on one particular hammer stroke: at the moment it hits the nailhead, the hammer has kinetic energy, KE; the amount of KE—the number of joules—depends on the weight of the hammer and its speed at the moment of impact. On hitting the nail, the hammer is brought to an abrupt stop; most of its KE is passed on to the nail, which penetrates the wood until friction stops it. The remaining energy is converted to other forms; some of it becomes waste heat at the point of contact of hammer and nailhead.

The rest of the energy sets both the hammer and the nail vibrating, making the air in contact with them vibrate too, so generating a sound. The vibrations start in the molecules of air touching the metal surfaces, which dislodge the molecules adjacent to them, which then dislodge molecules beyond them, and so on. In their vibrations, the molecules shift back and forth parallel with the direction in which the sound is traveling. Their movements are lengthwise (longitudinal), so sound waves are often called *longitudinal waves*. The vibrating air becomes successively compressed and rarefied, in layers that move steadily farther from the point where the hammer hit the nail. The vibrations reach the ears of anybody within range; rapid pressure changes in the air touching the eardrums are perceived as sound. In this way some of the energy of the original hammer stroke is carried far from its source while being partially absorbed, here and there, by solid objects it happens to touch.

Of course, the sound won't travel forever. As well as being absorbed by objects in their path, the vibrations of the air become attenuated as they spread out. At the same time, besides shifting back and forth through infinitesimal distances as they vibrate, the air molecules are also in a state of constant random motion in all directions, whose mean free path depends on the temperature (see chapter 3). Not surprisingly, repeated collisions among the air molecules ensure that pressure contrasts gradually become evened out between the compressed and rarefied layers, whereupon the sound fades away. All the original energy in the hammer stroke has now been dissipated: it has all become entropy.

A sudden sharp bang like the sound of a hammer stroke is difficult to analyze. Continuous sounds, like the roar of a waterfall or the hum of a quiet motor, are easier to deal with. The simplest sound of all is a prolonged pure musical note. Any continuous noise can be analyzed into a large number of component pure notes—or pure sounds as we'll call them—in the same way that the profile of a stormy sea can be analyzed into a large number of simple waves (see chapter 7). Recall that the energy of simple waves at sea is proportional to the square of the waves' heights. Analogously, the energy of any noise is proportional to the squares of the "heights" of all its component pure sound waves added together. What is meant by the height of a sound wave?

Consider figure 17.1, which shows two ways of portraying sound waves. Figure 17.1a shows sound pictorially: the varying density of the stippling represents the varying density of a representative "slice" of vibrating air; the sound is moving from left to right. Figure 17.1b translates the picture into the form of a wavy line showing how the air pressure varies. The speed at which the molecules move depends on the pressure differences, represented by the waves' height (the vertical distance from crests to troughs). To say that the energy of the waves is proportional to the square of their heights is therefore equivalent to saying that it is proportional to the square of the molecules' speed—which, when you recall how kinetic energy is defined (see chapter 3), is as it should be.[5]

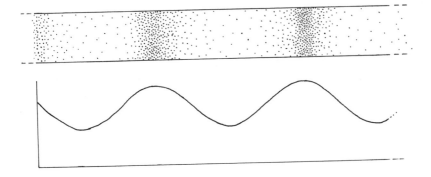

Figure 17.1. Two representations of the same series of sound waves. (a) The density of the stippling represents the density (hence the pressure) of the air. (b) The curve shows how air pressure rises and falls.

The total energy of a continuous sound accumulates as the sound goes on and on, whereas its loudness at any moment depends on the *rate* at which sound energy is being produced—on its power. As always, power is measured in watts (recall that one watt equals one joule per second). The *intensity* of a sound is defined as the power of the sound traveling across one square meter at right angles to its path. The units of intensity are watts per square meter, or in symbols, W m$^{-2}$.

It turns out that the human ear is astonishingly sensitive when you visualize watts in terms of a lightbulb's output. The quietest sound that the average human can hear—the threshold of hearing—has an intensity of $1 \times 10^{-12}$ W m$^{-2}$ (that is, one-trillionth of a watt per square meter). The intensity at which noise becomes painful is one trillion times as great, or 1 W m$^{-2}$, equivalent to the sound of a loud indoor rock concert.[6]

A more convenient measure of intensity has been devised to allow for the fact that, for the listener, a change in the intensity of a quiet sound is much more noticeable than an equivalent change in a loud sound. The *decibel* scale of loudness corrects for this.[7] The scale assigns a score of zero decibels (0 dB) to the threshold-of-hearing intensity and goes up from there, well past 120 dB, the threshold of pain. The rule is: If one sound is ten times more intense than another, then it is 10 dB louder; if it is $10^2$ (or 100) times more intense, then it is 20 dB louder; if it is $10^3$ (or 1,000) times more intense, then it is 30 dB louder; and so on.[8] As an example, the loudness of ordinary speech is about 60 dB. This means that talk is a million times more intense than the softest perceptible sound.

Thus far we have concentrated on sound waves traveling through air, which they do at a speed of 0.343 kilometers per second. They also travel through liquids and solids, but at much higher speeds. In fresh water the speed is 1.48 km per second; in granite, it is 6 km per second.[9] Converted to kilometers per hour (km/h) the speeds are: in air, 1,235 km/h; in fresh water, 5,328 km/h; in granite, 21,600 km/h. The more dense and the more rigid the medium, the higher the speed.[10] Note that the speed of traveling waves of sound is *not* the same as the speed of the minute back-and-forth movements of the molecules creating the pressure changes within each wave.

Very low frequency sound waves of great amplitude travel through the body of the earth as one of the varieties of seismic waves.

*Seismic Waves*

Most of an earthquake's energy is dissipated at the site of the quake. The rest is carried away in seismic waves. Like all traveling waves, seismic waves convey energy from one place to another. They are set in motion by an earthquake and travel outward, to be dissipated throughout a volume of the earth's interior and over an expanse of its surface: the energy is eventually converted to entropy and the waves die away. Let's look at the details.

Some seismic waves are giant sound waves. By analogy with the hammer-and-nail example of sound generation, an earthquake shock corresponds to a collision between a hammer and a nailhead, and the emitted seismic waves correspond to the emitted sound waves. Unlike sound waves, however, seismic waves are not all alike; on the contrary, there are several varieties. There are *surface waves*, generated most efficiently by shallow quakes, with a focus less than 70 km deep; they stay close to the surface, and they cause nearly all the damage. About three out of four quakes are shallow. Deeper quakes generate *body waves*, which spread in all directions, some passing right through the earth's center. Body waves are of two kinds: *primary waves* or *P-waves,* and *secondary waves* or *S-waves.* This gives us three kinds of waves to consider or, what comes to the same thing, three mechanisms by which solids and highly compressed liquids can convey energy across a distance.

First, consider the two kinds of body waves. Primary waves are so called because their speed is greatest and they arrive at a distant site first; they are also

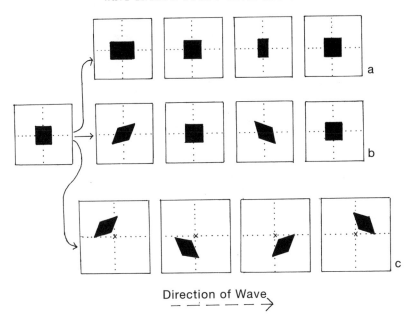

Direction of Wave

Figure 17.2. The single box on the left shows a cross section of a seismically "quiet" rock mass, just before seismic waves reach it; a fragment within the mass is colored black to show its changing shape as seismic waves pass. Three possible behaviors are shown, representing three types of waves; (a) a P-wave; (b) an S-wave; (c) an R-wave. (The distortions are exaggerated for clarity.)

called pressure or push-pull (compression) waves. Secondary waves travel more slowly; they come in second; they are also called shear waves. (Both kinds of waves have other descriptive names too, but they're not alliterative.) The two kinds of body waves operate in radically different ways.

P-waves are very low frequency sound waves; they entail the alternate expansion and contraction of each of the tiny fragments of rock that make up a large rock mass; the rock changes volume rhythmically, without any shearing. The reverse is true of S-waves: each fragment of rock changes shape rhythmically because of shearing, while its volume remains the same.

What happens is illustrated in figure 17.2. Each box in the figure represents the same vertical cross section of massive underground rock. The isolated box on the left shows the rock mass while the earth is "quiet," before the first jolt of an earthquake. One particular block of rock, though continuous with its surroundings, has been singled out by coloring it black. Each of the three rows of boxes on the right shows how the block behaves as a seismic wave travels through the whole rock mass. The boxes in each row show the same scene at a sequence of times. Rows a and b show the two kinds of body waves, P-waves and S-waves respectively. Row c shows a surface wave, also known as a *Rayleigh wave*, or R-wave.[11]

In R-waves, the block we are concentrating on changes in both shape and location. Because the waves are shallow and the rock is less dense than at greater depth, the particle motion (all in the vertical plane) causes appreciable ground movement. The waves, aptly described as ground roll, resemble a swell at sea, but with a surprising difference. The rock particles, which trace out vertical ellipses, move backward relative to the direction of the wave: if the wave is going from left to right (as in the figure) the particles move counterclockwise, and vice versa. Compare this with the behavior of a particle of water circling within a water wave (see figs. 7.3 and 7.7).

Now to compare the speeds and frequencies of seismic waves with those of sound waves. (Note that we are not concerned with the tiny distances traversed by molecules and particles *within* the medium, which are much less in liquids, and very much less in solids, than they are in air. Here we are considering the speeds and sizes of entire waves.) We have already compared the speed of longitudinal waves in rock and in air: recall that in air, sound waves have a speed of 0.343 kilometers per second, whereas in rock, where the longitudinal seismic waves are P-waves, their speed near the surface averages about 6 km/s or more than seventeen times as fast. S-waves and R-waves are slightly slower than P-waves, with speeds of about 3.5 and 3.1 km/s, respectively. All these seismic wave speeds depend on the type of rock that the waves travel through, and they increase at progressively greater depths below the surface because the pressure increases.

The frequencies of sound waves determine their pitch; the deepest note the average person can hear is about twenty waves per second. The frequencies of seismic waves are much lower—far too low to be heard (the rumble of a quake is ordinary sound, a by-product of the quake). For P-waves, S-waves, and R-waves, only a fraction of a wave passes per second: the average frequencies are 0.1, 0.06, and 0.04 waves per second, respectively.[12] These are averages: the output of a quake never consists of "pure notes." Rather, the waves of each type cover a band of frequencies, and low-frequency waves, which travel slightly faster than high-frequency ones, gradually pull ahead; this matches what happens to ocean waves when the whitecaps of a big "sea" travel away from the storm that caused them and become spaced out as an ocean swell (see fig. 7.5).

To make a long story short, there are always a variety of movements all happening simultaneously in the earth when seismic waves pass through. Measuring the energy output of a quake is correspondingly complicated. Exact and precisely timed measurements of earth movements (made with a *seismometer*) are unlikely to be obtainable anywhere near the epicenter of a big quake because seismometers there are likely to be broken, and even if they are not, they may give unreliable readings. The energy of a quake therefore has to be deduced from observations made far from the epicenter, in different directions and at different distances. A long list of factors has to be allowed for in the computations.

To begin with, the depth of the focus below the ground surface needs to be estimated. This is less than 100 km in the great majority of quakes and probably never exceeds 700 km, the depth at which the rock of the earth's mantle abruptly becomes more viscous and some of the subducting plates stop sinking (see chapter 15).[13] To judge the magnitude of a quake, a seismologist then needs to discover the distance from each seismometer to the quake's focus, a task fraught with difficulties. Seismic waves do not travel in straight lines: they speed up as they descend, because of the increasing pressure; this causes their paths through the mantle to curve and to be deflected sharply if they reach the core. P-waves can continue through the core, but S-waves come to a halt as soon as they reach it, because the outer core is liquid. It is easy to see why P-waves travel through both solids and fluids (liquids and gases), whereas S-waves can travel only through solids. P-waves are compression waves, and fluids as well as (most) solids, when compressed, tend to spring back when the pressure is relaxed: that's what makes the waves. S-waves, however, are shear waves, and fluids do not spring back when a deforming (shearing) force is removed: they are limp.[14]

To sum up: Quakes produce a mixture of seismic waves of different types, with varying amplitudes and frequencies; the waves follow various paths through rocks of different densities, at a range of pressures. Discovering the extent of earth movement at the focus of a quake (the quake's *magnitude*) is unavoidably complicated. After making allowances for all the complexities, however, magnitudes can be computed approximately. For a big quake, the results are publicly announced as being "on the Richter scale" in honor of the American seismologist who, in 1935, first devised a formula for measuring earthquake magnitudes.[15] His original formula has been improved and up-dated several times. Here we leap over all the difficulties, noting only that when magnitudes are measured at different observing stations, the results, though close, aren't necessarily identical. Moreover, the magnitudes of surface waves and body waves are seldom kept separate in newspaper accounts of an earthquake, although the difference is usually appreciable. For example, the British Geological Survey reported two magnitudes for the great earthquake that devastated Izmit, Turkey, and neighboring cities in August 1999: the magnitude of the body waves was 6.87, compared with 7.5 for the much more destructive surface waves.[16] The magnitude scale for quakes behaves like the decibel scale for sounds: an increase of one on the magnitude scale for quakes indicates a tenfold increase in the extent of earth movement, just as an increase of one on the decibel scale for sound indicates a tenfold increase in sound intensity. Discovering the energy liberated by a quake requires yet another step beyond computing its magnitude. It turns out that if a quake has a magnitude that is one unit greater than that of another quake, then (provided both are greater than magnitude 5) the stronger quake liberates between twenty-seven and twenty-eight *times* as much energy as the weaker.[17] For example, a quake of magnitude 6 liberates about $7.6 \times 10^{10}$ kJ, and one of magnitude 7 liberates about $2.1 \times 10^{12}$ kJ; in this case the energy of the stronger one is 27.6 times that of the weaker.

Estimates show that, over a study period of sixteen years (1977 to 1993) seismic waves transported earthquake energy through the earth at an average rate of about 4.7 million kilowatts.[18] This represents the combined power of the waves emanating continually from earthquakes of all intensities happening in all the world's earthquake zones. The energy radiated by seismic waves is only about one-twentieth of the total energy produced by earthquakes, however.[19] As remarked earlier, seismic waves radiating from a quake are analogous to sound waves radiating from the stroke of a hammer on a nailhead: they carry away only a small part of the energy generated. Seismic waves are, metaphorically, the background noise of the shifting tectonic plates, an inaudible noise, but one going on under our feet, in the earth's interior, all the time. They are carrying away the leftover energy of all earthquakes, after the initial, largest waves have done their damage if the quake was big.

# 18 WAVE ENERGY

## ELECTROMAGNETIC WAVES

*Waves without a Medium*

The waves we considered in the preceding chapter consist of regular, periodic disturbances that carry energy farther and farther away from its source. Exactly the same can be said of electromagnetic waves, the carriers of *radiant energy*. Electromagnetic waves (hereafter EM waves) differ profoundly from other kinds of waves, however, in that they need no material medium—they can carry energy through a vacuum. Seismic waves obviously need a medium; they are disturbances that form in and travel through the solid earth and the liquid iron of the earth's outer core. Sound waves are disturbances that form in and travel through many materials, including gases, liquids, and solids; that they cannot travel through empty space is not immediately obvious to the uninitiated, but most people remember from their school days how the sound of a ringing alarm clock enclosed in an airtight jar fades to silence when the air in the jar is pumped out.

EM waves, by contrast, can travel through empty space. They are regular, periodic disturbances in electric and magnetic fields in space. Admittedly, most EM waves can also travel through some material substances, with the substance acting as no more than a penetrable barrier. But when you look at the stars on a clear, dark night in the country, it is apparent that light travels best, and farthest, through interstellar space—empty but for sparse interstellar dust.

Before attempting to answer the question, What are EM waves? it's necessary to realize that EM waves are not the only possible representation, or model, of what radiant energy (including light) "really" is. EM waves are the so-called classical model. Modern particle physicists prefer to regard radiant energy as a stream of particles of zero mass, known as *photons*, about which more below.

The existence of different models does not mean, however, that one of them must be wrong. On the contrary, both are simultaneously right and wrong. They are right in the sense that both provide useful mental images of how radiant energy works. The EM-wave model is the most widely known and is almost entirely adequate for comprehending radiant energy on a "large," or "human," scale; the one case in which it is not—the photoelectric effect—is described in a later section. The photon model (in other words, the quantum theory model: a photon is a quantum of light) is needed for comprehending the subject on a subatomic scale. Both models are right in the sense that they provide adequate explanations of the phenomena they set out to explain. Both models could be called wrong in the sense that they certainly provide an incomplete picture. Scientific discoveries will undoubtedly continue for as long as our species, or at any rate our current civilization, persists and for as long as scientists can afford to experiment and observe. It is naive to suppose that there will be no new, unimagined phenomena to detect and no new problems to tackle. When the need arises, ingenious new models will be devised, or new mental images will be envisaged, to explain the new knowledge temporarily. Science will always be incomplete. There is no reason to believe that humans as a species have the capacity to understand everything there is to understand or that our thoughts provide anything more profound than mental images.

That said, we return to EM waves—what they are, how they are generated, and how their energy is dissipated.

## The Nature of Electromagnetic Waves

As we saw in chapter 16, a magnetic field appears spontaneously in the neighborhood of an electric current; and a current flows spontaneously in a conduc-

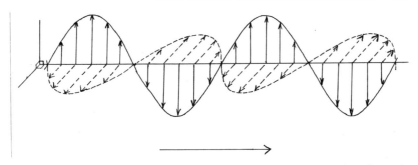

Figure 18.1. Electromagnetic waves. The solid arrows in the vertical plane show the varying strength and direction of the electric field; the dashed arrows in the horizontal plane show the varying strength and direction of the magnetic field. (For each field, the longer the arrow, the stronger the field.) The waves are moving from left to right.

tor moving through a magnetic field. Recall, too, that a current consists of moving electric charges, and moving charges create moving electric fields. Putting these facts together, we realize that moving electric fields create magnetic fields and vice versa: moving magnetic fields create electric fields. The processes are symmetrical. And because the two kinds of fields are absolutely dependent on each other, they can be thought of jointly as the two components of a single electromagnetic field.

Electromagnetic waves are moving "disturbances" in an electromagnetic field, caused when moving electrons accelerate or decelerate. Consider what would happen if you caused an electric field to *oscillate* or, what comes to the same thing, vibrate. Bear in mind that an electric field has both a direction (see fig. 16.3) and a magnitude or strength, and the same goes for a magnetic field. An oscillating field is one whose strength grows from zero to a maximum pointing in one direction, say to the right, then dwindles to zero again, then grows to the same maximum pointing to the left, then dwindles to zero again, then . . . And so on. It never stops accelerating, either to the left or to the right. Now imagine that an electric field starts to oscillate; its coupled magnetic field is forced to oscillate too. The magnetic field's strength grows and dwindles in time with the electric field's, but at right angles to it. The result is a train of EM waves, represented pictorially in figure 18.1.

The figure shows waves, but they are conceptual waves, not physical ones: there are no actual ripples on a surface. The wavy lines simply join the tips of successive arrows drawn to show the instantaneous direction and strength of the coupled electric and magnetic fields. The figure can be interpreted in two ways: either as an instant snapshot of a constantly changing pattern or as a record of the way the field strengths change with passing time at a fixed location. (Recall that the same is true of waves in a material medium; see chapter 17.)

Now for the energy transported by the waves. It is customarily measured as power, in watts, so to calculate the number of joules transported in a given time interval one must multiply the wattage by the number of seconds. The power at any instant is proportional to the magnitude of the electric field ($E$) multiplied by the magnitude of the magnetic field ($B$) at the same instant.[1] Instantaneous power varies rapidly, from zero to a maximum and back again; practical measuring instruments (and human eyes in the case of light waves) automatically average the instantaneous power over thousands of "instants" to give a measurement of power in the ordinary sense, that is, sustained power.

Electromagnetic waves bear energy through empty space at the enormous speed of 300,000 kilometers per second, or in scientific notation, $3 \times 10^8$ m s$^{-1}$. This is one of the most famous numbers in all science; it is universally represented by the letter $c$, always in lowercase and always italicized. This is the $c$ of Einstein's famous equation $E = mc^2$. It never varies, because it is determined by the electrical and magnetic characteristics of space, which are unchanging. No material particle, however small, can go as fast as or faster than $c$; only EM waves can attain this greatest of all possible speeds, and then only in a true vacuum. They travel at somewhat lesser speeds in a material medium; for instance, at about $2.3 \times 10^8$ m s$^{-1}$ in water and $2 \times 10^8$ m s$^{-1}$ in glass.

The mathematical equations relating the speed, $c$, to the intrinsic characteristics of space were discovered in 1864 by the Scottish physicist James Clark Maxwell (1831-79), and it was he who first computed the numerical value of $c$. He immediately realized that it was the same as the speed of light, which had been accurately measured not long before. The precise equality of these two speeds led Maxwell to realize that light and all other forms of radiant energy consist of EM waves. Like the discoveries of his predecessor Newton in the seventeenth century and his successor Einstein in the twentieth, Maxwell's discovery advanced human knowledge of the physical world enormously. It was the most important achievement of nineteenth-century physical science.

## How Radiant Energy Is Generated

Every source of radiant energy, or equivalently, every generator of EM waves whether natural or artificial, must cause electrical charges to accelerate. It hap-

pens, or is made to happen, in various ways: sometimes the electrons accelerate sporadically, sometimes they oscillate. Although all EM waves are alike in
consisting of disturbances in the electromagnetic field, they are not all alike in
their wavelengths (or conversely, their frequencies). In discussing the different kinds of waves, they can be labeled by either their wavelengths or their
frequencies, and here we use wavelengths. It's simple to convert from one to
the other, using the formula

$$\text{wavelength} \times \text{frequency} = c.$$

The range of wavelengths to consider is huge, from about one-trillionth of
a millimeter to over 5,000 kilometers; theoretically there is no upper limit. Besides differing in wavelength, EM waves also differ in the energy they contain:
the shorter the wavelength the greater the energy.

This statement seems surprising at first. It appears to contradict the asser-
tion that the power of EM radiation depends on the intensities of the electromagnetic fields, that is, on the amplitudes of the "waves" in figure 18.1. When
EM waves are produced artificially, why can't their energy be increased at will
simply by increasing the energy input? On a human scale, this can easily be
done. But the statement that shorter wavelengths carry more energy than
longer ones refers to the minute, indivisible *quanta* of which radiant energy
consists.

A quantum of radiant energy (a photon) is a single "grain" of energy in the
same way that an atom is a single grain of matter, except that it has zero mass.
In spite of this, a photon does have energy. Einstein, in 1905, hypothesized that
the energy, $E$, of a single photon in a vacuum depends on the wavelength, $L$,
of the radiating waves thus:[2]

$$E = hc/L \text{ joules.}$$

For a given wavelength, no smaller unit of radiant energy can exist. The constant $h$ is called Planck's constant, after the German  physicist Max Planck
(1858-1947), who discovered it in another context. Numerically, $h$ is mind-
bogglingly small; it is $6.63 \times 10^{-34}$ joule-seconds; $h$ is as central to modern
quantum physics as $c$ is to relativity theory.

Let's compare the energies of some different kinds of photons. They are
easily calculated once you know the wavelengths of the radiation concerned.
The energies are best measured in electronvolts (eV) rather than joules (J), because they are so small (recall, from chapter 10, that 1 eV = $1.6 \times 10^{-19}$ J).
Some examples, in round numbers:

A photon of hard X rays of wavelength $L = 10^{-11}$ m has energy of 100,000 eV.

A photon of red light of wavelength $L = 7.5 \times 10^{-7}$ m has energy of 1.7 eV.

A photon of broadcast radio waves with $L = 100$ m has energy
of 0.00000001 eV, and so on.

The energies are given here in "ordinary" numbers (rather than in scientific notation) to emphasize their enormous range. Although the energy in a single photon is always tiny, a photon of "hard" (meaning short) X rays is about ten trillion ($10^{13}$) times as energetic as a photon from a radio station's transmitter.

Not surprisingly, EM waves of widely different wavelengths are generated by different processes, in nature as well as in artificial settings—laboratories, factories, houses, and the like. The most energetic radiation is γ-radiation (γ is the Greek letter gamma), which is emitted when certain radioactive nuclei undergo γ-decay; the decaying nuclei lose energy while their masses remain unchanged. There is no conversion of mass into energy in γ-decay, as there is in α-decay (described in chapter 13); the energy comes from an "excited" atomic nucleus—one with surplus energy—as it returns to its unexcited, ground state. The wavelength of the rays is comparable (in ballpark terms) to the diameter of a nucleus, about $10^{-15}$ m. Gamma rays are emitted, along with damaging subatomic particles, by nuclear weapons, faulty nuclear power plants, and radioactive material stored as nuclear fuel; the dangers they pose are known to everybody. Gamma rays also come, in absolutely harmless amounts, from the sun and a variety of other astronomical sources, some immensely distant. They are also emitted harmlessly by naturally occurring radioactive elements such as uranium and thorium, which are present at very low concentrations in the rocks, particularly in granite; all emit γ-radiation in negligible amounts.

Below γ-rays in terms of photon energy, and with longer wavelengths, come X rays; their wavelengths are in the neighborhood of $10^{-10}$ m. They are produced artificially, by bombarding a metal "target" with a stream of fast-moving electrons. On hitting the target, the electrons are brought to an abrupt stop. They are sharply decelerated, or "braked." The kinetic energy lost by the electrons becomes the energy of EM waves, which are emitted by the target; X rays are a form of this so-called braking radiation.[3]

Longer wavelengths are produced when the electrons inside atoms move. As we saw in chapter 13, most of an atom's volume consists of the empty space surrounding its nucleus, and all this space is available to its electrons. The diameter of the space depends on the element an atom belongs to: $10^{-10}$ m is a representative figure. An electron in this space can occupy any of several distinct energy levels, and if it "jumps" from a higher level to a lower, it emits an amount of radiant energy equal to the "height" of the jump (the difference between the two energy levels). The jump is known as an *electronic transition.*[4]

The EM waves emitted when electrons perform such transitions belong to several familiar wave bands: in order of decreasing photon energy they are ultraviolet rays and visible light, with wavelengths ranging from about $10^{-8}$ m up to nearly $10^{-6}$ m (or 1 μm), and also some longer (infrared) waves. The electrons are first boosted from lower to higher energy levels by energy from an outside source, and then they reemit the energy by dropping back down again.

Ultraviolet light is artificially produced by causing an electric current to flow through a gas, such as hydrogen or mercury vapor, in a sealed tube: the flowing electrons cause electronic transitions in the atoms of the gas. The same process is used to produce some kinds of visible light, for example, outdoors in neon tubes and indoors in fluorescent lights.

*Incandescence,* the emission of light by objects that are red-hot or white-hot, is the source of most visible light; not surprisingly, light produced in this way is always accompanied by radiant heat. Electronic transitions produce all of the light and some of the heat. Most of the heat is generated by a different process, however, as we see below.

Almost all the natural light on earth comes from the surface of the sun, heated to incandescence by the nuclear reactions in its interior. Indeed, light is emitted by any object heated to a high enough temperature, whether it be the sun, or a glowing ember left after a fire, or the filament of an incandescent lightbulb; the filament heats up because of collisions between the flowing electrons of the current and the particles of the conductor.

Visible light consists of EM waves in the range $0.4 \times 10^{-6}$ m (violet) to $0.7 \times 10^{-6}$ m (red), as we saw in figure 11.1. Longer waves, because they are invisible, are labeled "infrared," but the contrast between visible waves and infrared ones is more in the sensations they produce than in the way they are generated. With light waves, the different wavelengths are seen as different colors; with infrared the only sensation is warmth, whatever the wavelength. It is only an accident of human evolution that our skins don't perceive something analogous to colors corresponding to different infrared wavelengths. It would be fascinating if short wave and long wave infrared felt as different from each other as the colors blue and orange look.

Infrared radiation by itself is emitted by heated objects not hot enough to glow. Some of the radiation comes from electronic transitions, but most comes from the "shaking" that electrons undergo as passengers on molecules in constant movement.[5] Many molecules vibrate and rotate, and all participate in the constant random motion that constitutes thermal energy. Indeed, thermal (infrared) radiation is emitted all the time, by *every* object whose temperature is above absolute zero. Recall that in chapter 4 we considered the heat budget of the whole earth and the way the ground reradiates heat from the sun (see fig. 4.1). Similar heat exchanges are always happening on a smaller scale, everywhere; for instance, all the surfaces in a room—floor, ceiling, walls, furniture, people's skins—are constantly radiating, absorbing, and reradiating infrared radiation.

We now come to a narrow waveband centered on a wavelength of about 1 mm. It isn't usually listed as a labeled band, and the waves aren't (except incidentally) generated artificially to serve a useful purpose. They are the EM waves of the *cosmic background radiation* that is believed to permeate all space and to be the surviving energy of the Big Bang, now thinly spread through the enormously expanded (and still expanding) universe.[6]

Artificially generated radiation with wavelengths from 1 mm to 30 cm (bracketing the cosmic background radiation) is called microwave radiation. It is used in radar and for microwave cooking.

To conclude this list of wave bands, let's consider some of the longest waves assigned to a labeled band—radio waves. They range in length from about 10 cm to 100 km; wherever there's a radio or television transmitter, they are artificially generated all day long and sent forth over the "airwaves." The heart of any transmitter is an electric *oscillator*, a circuit so designed that the current in it constantly oscillates at a chosen frequency. The magnetic field around the current-carrying wire automatically oscillates in unison. The oscillating magnetic field creates its own oscillating electric field, which augments the oscillating magnetic field, which augments the oscillating electric field, which . . . . And so on. In a word, EM waves are generated. The two oscillating fields can be said to "feed off each other."[7]

Very long radio waves are transmitted coincidentally wherever alternating currents flow. Ordinary domestic alternating current (AC) alternates—oscillates—at a frequency of sixty cycles per second, or 60 hertz. The resultant EM waves are 5,000 km long and have a photon energy about one–ten trillionth that of red light. Theoretically, there is no upper limit to the wavelength of an EM wave. If you wave a garment with static cling back and forth, then some very, very long EM waves (of immeasurably small energy) will be emitted.

Strong radio waves are produced in the natural, outdoor world by lightning flashes; the sudden flow of electrons from one part of a thundercloud to another, or from the cloud base to the ground (see fig. 16.4), sends out powerful pulses of radio waves, heard as loud crackling if you turn a radio on during an electrical storm. Lightning is the only appreciable source of natural radio waves on earth, but extraterrestrial radio sources, including "radio stars," are known to be numerous.

It's worth reemphasizing that all the wavelengths we have considered are present in solar radiation, although not in equal proportions; the biggest proportion of solar energy comes as visible light and "near" ultraviolet light. It is not surprising that human eyes have evolved to be sensitive to a wave band that almost coincides with the strongest segment of the solar spectrum (see fig. 11.1), but why isn't the coincidence total? It is a mystery why we cannot see near ultraviolet radiation, which is as strong a component of sunlight as visible red light; bees can see it.

## How Radiant Energy Is Dissipated

The moment EM waves are generated, they start traveling away from their source at 300,000 km per second, bearing energy. Where do they go, and what becomes of the energy?

The most energetic of them, γ-rays and X rays, carry so much energy that it takes relatively few photons to cause injury and death to living things (including people) exposed to them. Their energy is transferred directly to molecules of living tissue, causing injurious chemical changes including burns. The photons of these "hazardous" radiations are energetic enough to dislodge electrons from atoms.[8] Even waves as long as light waves can shift the electrons in metals, causing a current to flow; comparatively little energy is needed for this, because the electrons in metals "roam" free and unattached (see chapter 16). The phenomenon is known as the *photoelectric effect*; it is the power source for solar-powered pocket calculators, photographers' light meters, burglar alarm systems, automatic door openers, and the like.

The photoelectric effect doesn't seem especially noteworthy on first acquaintance, but investigation of the details led to a profound scientific advance—the realization that EM waves must consist of "grains" of energy (photons), as we saw above. This discovery, not the theory of general relativity, earned Einstein his Nobel Prize.

Two characteristics of the photoelectric effect underlay the discovery. First, for a current to flow in an irradiated metal, the wavelength of the radiation must be shorter than a certain threshold length that depends on the metal; for example, if it is a pellet of the metal sodium, the radiation must be blue-green or bluer; if it is a copper wire, then only ultraviolet will suffice. Second, provided the wavelength is short enough to cause a current at all, then increasing the intensity of the light increases the current.

These two facts imply that radiation must consist of discrete photons, each a separate package of energy. Experiments show that to cause any current at all to flow in sodium, each photon must carry at least 2.3 eV of energy; the corresponding threshold energy for copper is 4.7 eV. No matter how numerous the photons (no matter how intense the light), if the photons are individually too weak nothing happens: you can't break a window by pelting it with feathers, no matter how numerous the feathers. But if the individual photons are energetic enough, (if their wavelengths are short enough), then the more of them there are—the more intense the radiation—the greater the photoelectric current.

To sum up, when high-energy EM waves strike any matter, or when waves with lower energy strike "susceptible" metals, they dislodge electrons; the energy of each photon of radiation is converted to the kinetic energy of an electron.

Waves of ultraviolet and visible light, though less energetic than $\gamma$-rays and X rays, are powerful enough to rearrange the internal structures of atoms and molecules by breaking chemical bonds; that is, they bring about *photochemical* reactions.

Some examples: The ultraviolet rays in sunlight cause chemical changes in human skin cells. The results can be anything from gratifying (a good tan) to unpleasant (painful sunburn) to life threatening (some skin cancers). The bark of trees can also be severely injured by sunburn.[9]

When bright sunlight shines on city air polluted with nitrogen dioxide, the ultraviolet component of the light energizes a photochemical reaction, producing smog.[10]

Whenever you take a photo, a photochemical reaction changes every molecule at the surface of the exposed film emulsion.

But these are trivial examples of photochemical reactions. The single most important chemical reaction of any kind, from the point of view of (nearly) all life on earth, is photosynthesis, described in chapter 11. Directly or indirectly, it creates virtually all living matter. Looked at from another angle, it is one of the processes in which electromagnetic energy from the sun is consumed: it is transformed into biochemical energy.

Electromagnetic waves are also consumed, in less spectacular fashion, whenever their energy is transformed into heat directly, by speeding up the random motions of atoms and molecules. The waves are said to be absorbed. Absorption is what happens to most infrared radiation. It goes on everywhere, whenever an object is warmed by the sun, by a fire, or by radiation from anything warmer than itself.

The energy of radio waves emitted by transmitters is used, for the most part, in forcing oscillating electric currents of matching frequency to flow in receiving antennas (and incidentally, in anything else that conducts electricity). The energy that began as the kinetic energy of electrons in a transmitting antenna ends up as the kinetic energy of electrons in a distant receiver, having crossed the gap between transmitter and receiver as energy-bearing waves.

Electromagnetic waves are not automatically absorbed by whatever they strike. This is obviously true of light waves, as anybody can see by looking through clean, clear, colorless glass: it is as though the glass weren't there. In other words, glass is transparent. Other materials—sheet metal, for instance—are opaque; they absorb all the light that falls on them, letting none pass through.

This raises a problem: Why are some materials transparent and others opaque? The answer is a highly technical part of solid-state quantum physics; here it is possible to say only that it depends on the behavior of the electrons in the material, behavior that also affects its electrical conductivity. To generalize, metals are opaque because they are conductors; insulators are transparent because they are insulators.[11] For example, glass and clear plastics are simultaneously transparent and good insulators. But what about such seemingly opaque insulating material as china and nonclear plastics? In fact, these materials aren't truly opaque, in the sense that they absorb light. Rather, light shining on them is scattered by a myriad of structural irregularities in the material (more on scattering below).

Different materials are transparent to different wavelengths of radiation. For example, bone is opaque to X rays, but flesh is transparent to them. Flesh is opaque to infrared rays, however; if it were transparent to them, they could have no warming effect. Most glass is transparent to sunlight but opaque to ultraviolet; good sunglasses protect eyesight by blocking ultraviolet rays. Gases can be opaque too. As everybody knows these days, ozone is opaque to ultraviolet radiation; the thinning of the ozone layer high in the atmosphere because of air pollution is permitting energetic ultraviolet rays to reach ground level, where they damage  living organisms including humans. As is equally well known, carbon dioxide is opaque to infrared radiation. This is thought to be the principal cause of global warming: a growing proportion of the earth's annual heat income from sunlight is prevented from being reradiated skyward by ever-increasing concentrations in the atmosphere of gases opaque to infrared rays—the "greenhouse gases."

The solar energy budget was described in broad outline in chapter 4, and it was noted that, apart from the sunlight reflected by clouds, most of the rest is absorbed and reradiated. A much smaller fraction, too small to be shown in figure 4.1, remains to be accounted for; it too is reflected, but not by clouds. It is *scattered*. Light is scattered by everything it strikes. Here we consider scattering in the atmosphere; scattered sunlight from the atmosphere is as important for life, especially plants, as direct sunlight. It reaches the ground regardless of whether the sky is clear or cloudy. When it's cloudy, although much of the sunlight is reflected back to space by the tops of the clouds and some is absorbed by cloud droplets, scattering still goes on above, within, and below the clouds. Scattering causes negligible reduction in the energy of the affected light; rather, it spreads the light out over the whole sky.

Scattering is the reflection of light by a vast number of tiny reflectors separated by comparatively wide gaps. Alternatively, one can define reflection as a form of scattering: it is the "scattering of light by a large number of [closely spaced] scattering centers."[12] This amounts to saying that scattering and reflection are two forms of a single phenomenon. In some scattering, however, the tiny reflectors, or "scatterers," are no bigger than the wavelength of the light reflected, and instead of sending the radiation back toward its source, they *deflect* it in all directions. Some of it is even sent onward in its original direction because the energy of the light is absorbed by the scatterer and then immediately reemitted.

The way reemission happens depends on the sizes of the  scatterers; the smallest of them are individual molecules of the air's oxygen and nitrogen, with diameters less than one-thousandth of the wavelengths of light.[13] Slightly larger are tight clumps of air molecules, airborne bacterial spores, minute salt crystals evaporated from ocean spray, particles of fly ash from forest fires and industry, and the like. With particles of this size, the blue component of sunlight is deflected much more strongly than the red; hence the blue skies of a sunny day. When the particles are larger and the air is hazy with dust or misty with minute water droplets, most of the light is sent onward as white light, away from the source, producing a white glare in all directions, brightest toward the sun. When the particles are much wider than a wavelength—as large as falling raindrops, for instance—much of the sun's energy is reflected back into space without deflection: less light reaches the ground, and the cloudy sky becomes noticeably dark.

As for solar energy, the point to observe is that little of it is returned skyward on a clear day and scarcely more on hazy or misty days. Rather, it is scattered and spread around. Some reaches the ground from all directions, while that coming in a straight line from the sun itself is correspondingly reduced. Life on the surface, especially at low elevations, has adapted accordingly: "shade plants" obtain enough indirect light energy to photosynthesize; at the same time, desert life is able to survive the direct light.

Finally, what becomes of all the solar energy that bypasses the planets? The earth intercepts less than one-billionth of the radiation flowing out from the sun in all directions. The EM waves travel away into space, almost unhindered, becoming attenuated as they spread out. They inevitably lose a tiny amount of energy whenever they encounter a "body" of any kind, but this becomes apparent only when the body is a comet not too far from the sun. Then it is possible to see how *radiation pressure* imparts some solar energy to the cloud of minute ice crystals and fine dust that forms the comet's tail. The pressure is proportional to the number of joules of energy per cubic meter of space.[14] The wave-borne electromagnetic energy is converted to kinetic energy of the particles, causing them to stream away from the comet's head on the side away the sun. Both the big comets of the 1990s, Hayakutake in 1996 and Hale-Bopp in 1997, demonstrated the phenomenon splendidly. Comets' tails give us our only chance to observe what is probably the weakest manifestation of solar energy at work. Radiation pressure produces no discernible effect on earth.

# 19 HOW ENERGY IS USED

*Sources of Raw Energy*

Every living thing, without exception, depends on a source of energy outside itself to keep it living and reproducing. Sunlight provides the energy for green plants; animals use the chemical energy in their food. And all living things except for subterranean microbes are warmed, directly or indirectly, by the sun's radiant heat.[1] For all nonhuman animals, food and sunshine are the only sources of energy. This final chapter gives a brief overview of the multitude of energy sources tapped by humans and of some of the ways "raw" natural energy is converted for human use. The whole field of energy technology is advancing so rapidly at present that detailed accounts of all the new developments would be a vast undertaking and would soon be out of date; so would estimates of the quantities of energy obtainable from different sources, and the costs. It is worthwhile, however, to consider how the multitude of natural energy sources described in previous chapters are being used by our species.

The first animal species to succeed in acquiring energy besides that supplied by food and sunlight was probably Neanderthal man (*Homo neanderthalensis*); ashes and charcoal are often found with the fossil remains of Neanderthals, suggesting that they used fire regularly, presumably for warmth and perhaps for cooking too. The species is believed to have branched off from the direct line of descent leading to us (*Homo sapiens*) about 200,000 years ago and to have disappeared 170,000 years later.[2] In the Ice Age climate that prevailed in much of the Neanderthals' homelands—Europe and western Asia—during much of their time on earth, life without fire would have been impossible. The first fuel used for fires was presumably wood. Long after the Neanderthals had died out, leaving us as the only living species of humans, additional fuels came into use. Once humans had begun to domesticate grazing animals, between 5,000 and 10,000 years ago, their dried dung became an easily available fuel; cows, water buffalo, and yaks were probably the main contributors. The energy of the living animals themselves was also used: they served as beasts of burden.

Other sources of energy may have been in use even longer than combustible fuels. Geothermal energy was available and waiting to be exploited in some parts of the world, and no doubt the first humans to settle these regions quickly availed themselves of the unexpected luxury.

Other widespread energy sources that early humans must have exploited were "river power" for conveying the first rafts or boats and wind power for sailing them.

Some of these sources, particularly combustible fuels and the energy of moving fluids, have presumably been used for many thousands of years. They illustrate two of the ways energy can be obtained: first, by burning a combustible fuel to liberate its stored chemical energy and, second, by borrowing ready-made kinetic energy.

### Combustible Fuels

The combustible fuels available to primitive human societies were wood, peat, and dung, and in some localities modest amounts of coal, oil, and natural gas. Coal can be had without mining where it happens to outcrop at the surface. In a few places, for example, in the valleys of the Tigris and the Euphrates, oil seeps out of the ground unaided and has been used since prehistoric times.[3] Natural gas seeps out of the ground in parts of China and was in use as a fuel as early as 3,000 years ago.[4] Oil from large animals has been a fuel source for the indigenous population of Arctic Ocean shores for thousands of years; it was rendered from blubber, the thick layer of fat insulating the bodies of whales and seals living in cold seas.

All these fuels seem at first to be organic. Wood obviously is. Coal is the fossil remains of tropical swamp vegetation; it often contains easily recognizable fossils of some of the plants. Peat is the remains of fairly recently dead bog and swamp vegetation that has not yet fossilized to coal, though it will do so eventually if it happens to become deeply buried by overlying sediments and subjected to the higher temperatures and pressures of the depths. The plant remains that formed coal and peat did not decompose after death because they were submerged in water too acidic, or too short of oxygen, for decomposers to thrive. If the dead vegetation had decomposed, the chemical energy trapped in it would have been used up by the decomposers, and nothing would have been left to burn thousands or millions of years later.

Petroleum (oil and natural gas) consists of a mixture of hydrocarbons, chemicals composed wholly of hydrogen and carbon. It comes from the remains of dead plants and, at sea, from the millions of plankton organisms whose dead bodies rain down on the ocean floor unceasingly—not all of them are eaten by larger animals on the way down. Most petroleum originates beneath the ocean floor, and it usually contains small amounts of chemical compounds found elsewhere only in living, or once living, marine organisms.

If it were the whole story, the foregoing would lead to the conclusion that all the combustible fuels—wood, dung, peat, coal, animal oil, and petroleum—derive their stored chemical energy from organisms that, while alive, captured the energy of sunlight. But it is not the whole story. The astronomer and cosmologist Thomas Gold has recently advanced the hypothesis that much of the world's petroleum is inorganic in origin and started out as an ingredient of the primordial material of the solar system—the cloud of rock fragments, dust, and gas from which all the planets condensed (see chapter 14).[5] The giant gas planets, especially Jupiter and Saturn, are known to contain vast amounts of lightweight hydrocarbons, including the lightest of them all, methane (a methane molecule consists of one atom of carbon and four of hydrogen). Hydrocarbons frozen to ice have been detected on the surfaces of some of these planets' satellites. Given that hydrocarbons are abundant in the outer planets, it seems reasonable to suppose that the earth has its share of them too, in our case buried below the planet's surface. Supporting evidence is that some petroleum is found in deep igneous rocks rather than in marine sediments.

If the hypothesis is true (it is controversial at present) it follows that our traditional combustible fuels do not come entirely from living and fossil organic material, whose energy comes, indirectly, from sunlight. A fraction of the petroleum—how big a fraction is hard to judge—may be as ancient as the solar system itself and may never have been incorporated in living organisms. It seems unlikely that all petroleum belongs to the planet's primordial ingredients; the evidence that some of it comes from the remains of marine organisms remains as strong as ever. This leads to the tentative conclusion that not all petroleum has the same origin. It may be a far more interesting substance than previously realized.

In any case, the fossil fuels are treated as irreplaceable, although they would replace themselves given enough time—tens of millions of years. "Primordial petroleum" is certainly irreplaceable. If the earth's supply of these fuels continues to be consumed by humans at the current rate, it is likely to be exhausted long before our species goes extinct. Quickly renewable substitutes are urgently needed.

Among the substitutes coming into use are garbage gases recovered at landfill sites, methane from sewage treatment plants, biodiesel fuel from vegetable oils, and ethanol obtained by distilling agricultural crops, principally corn.

Potentially the most valuable substitute is hydrogen, even though its explosiveness makes it expensive to store and dangerous to use. It has the unique merit of not emitting greenhouse gases, which all the hydrocarbon fuels inevitably do; the only by-product when hydrogen is burned (that is, oxidized), is plain water. But because hydrogen doesn't occur pure and uncombined in nature—it is far too chemically active—it has to be extracted from compounds containing it. If these are hydrocarbons, as they usually are, the extraction process itself uses energy and emits pollutants, which greatly reduces the extracted hydrogen's value; the net gains in energy and cleanliness may be slight.

These two drawbacks have been overcome by using pure water as the raw material—hence no pollutants—and sunlight as the energy doing the extracting—hence no need for fossil fuels.[6] Photochemical reactions powered by sunlight split the water into hydrogen and oxygen; the oxygen is released to the atmosphere, and the hydrogen is collected and stored for later use as fuel. This is one of the ways of storing sunlight for later use.

Hydrogen can be used in two ways as an energy source: it can be burned as a combustible fuel, or it can made to combine nonexplosively with oxygen in such a way that the chemical energy liberated is converted directly to electricity. This is what happens in *fuel cells*.[7] Fuel cells can supply the energy for automobiles driven by electric motors. When they are the motive power for all vehicles, environmentally damaging emissions from vehicles will be a thing of the past. This doesn't mean, however, that other kinds of environmental damage will not continue to result from the use of road vehicles: but for them, it would not be necessary to cover vast areas of land with wide highways and huge parking lots that, together, obliterate existing ecosystems and derange the hydrology of underlying aquifers.

There may soon be a new addition to our list of combustible fuels.[8] It is *methane hydrate*, obtained as a frozen solid from the seafloor or from sediments directly below Arctic permafrost (perennially frozen ground). The frozen material is a form of ice in which the water molecules are linked into a three-dimensional network of cells ("cages") each containing a single molecule of methane. It was first found in permafrost areas in the 1960s and on the seafloor in 1970. Huge submarine deposits are now known to exist in many parts of the world; in some places hectares of the seafloor are carpeted with a thick white layer of it, and it also occurs in masses embedded in the uppermost few hundred meters of the seafloor sediments. The methane forming the hydrate presumably  comes from the same sources as natural gas: in part from long-dead organisms and in part from the earth's primordial material.

Methane hydrate remains frozen only where the temperature is low and the pressure high; it is found on the seabed at depths greater than five hundred meters. When it is lifted from the depths it becomes unstable: as the temperature rises and the pressure falls, the combustible methane bubbles off and only pure, liquid water remains. This will make "mining" it difficult and expensive until the needed technology is developed. The marine deposits of methane hydrate contain more carbon than all the other combustible fuels (wood, peat, and all the fossil fuels) combined, so it could be a veritable bonanza. But there may be a risk of liberating large quantities of unburned methane into the atmosphere as the hydrate is retrieved from the bottom of the ocean; this could cause an unwelcome speedup of global warming, because methane is a much more serious greenhouse gas than is carbon dioxide.

Let's return to combustible fuels in general. Burning them generates heat: in terms of energy, their stored chemical energy is converted, initially, to thermal energy. For Neanderthals this was probably enough. About 30,000 years later, in 1769, the Scottish inventor James Watt (1736-1819) invented the steam engine, and the heat of combustion was for the first time transformed into kinetic energy. It still is, in several ways. The heat from burning coal is sometimes used to boil water to make steam that, at high pressure, drives turbines or the pistons of external combustion engines. When gasoline, diesel oil, or propane burns in the cylinders of internal combustion engines, the forceful expansion of hot gases drives the pistons of land vehicles; similar engines drive motor boats and ships. In aircraft engines the expanding gases from burning fuel usually drive turbines. In a turboprop, the speed of the rapidly spinning turbines is adjusted to the speed of the propellers that make the plane fly. In a turbojet, the turbines power compressors, and the hot compressed gases then stream backward at high pressure, giving the plane tremendous forward thrust. In rockets a similar thrust is provided without the aid of compressors, by burning solid fuels. In all these processes, chemical energy in the fuel is liberated as thermal energy when the fuel burns, and the thermal energy then becomes kinetic energy; the KE may move land vehicles, ships, or aircraft or it may do its work in a stationary engine.

Another energy transformation takes place when the energy of spinning turbines, driven by combustible fuels (coal, oil, or gas), is converted to electricity in a generator; the sequence then has four steps: chemical energy to thermal energy to kinetic energy to electrical energy. Turbines can be made to spin by energy from other sources too, as we shall see below. All the same, combustible fuels are not likely to become obsolete as energy sources for some time to come.

## The Energy of the Wind

The kinetic energy in winds, waves, and rivers all comes directly from the radiant energy of the sun, and the energy is usable in a multitude of ways. Windmills and waterwheels have been converting the kinetic energy of moving fluids into mechanical energy to mill grain and lift loads for thousands of years. Windmills, after becoming almost obsolete by the mid-twentieth century, are now coming back into use to supply nonpolluting energy. They enable the kinetic energy of the wind to be converted into electrical energy on the spot, using generators driven by the windmills' rotors.

Modern wind farms cover hundreds of hectares with scores of tall windmills. The commonest kind has a rotor resembling an aircraft propeller rotating on a horizontal axis. Less common are "eggbeater" windmills that rotate on a vertical axis. A propeller-type rotor in an average wind spins at about three hundred revolutions per minute; this has to be geared up to a rotation rate ten times as great to drive the turbines of an electrical generator. The more uniform the rotation rate the better, so devices are used to compensate for variations in wind speed: the pitch of the rotors can be altered or the blades can be equipped with adjustable flaps; if the wind is too strong, the whole windmill can be rotated on its tower so that it does not face directly into the wind. Technical improvements continue to be made. Windmills are available for individual homes; the electrical energy they generate is usually stored in lead-acid batteries for use on demand. Here is an energy sequence starting with nuclear fusion (in the sun), via thermal energy driving the wind, to the kinetic energy of rotating windmill blades, to electrical energy produced by the windmill's generator, to chemical energy in the storage batteries, and finally back to electrical energy again to operate domestic appliances. Some energy is inevitably wasted at every stage; it becomes entropy.

## The Energy in Moving Water

Running water has been used as a source of energy since the dawn of history. It turns waterwheels. The modern form of waterwheel—now called a turbine—is the device that drives hydroelectric generators, providing a large fraction of the total energy used in industrialized countries. To increase the supply of running water and make it controllable, big dams have been built that create capacious reservoirs with enough head to yield an uninterrupted supply of fast-flowing water in large volumes. Dam building was one of the principal engineering enterprises of the twentieth century; it has completely changed the landscape in many parts of the world, particularly in western North America.[9]

Hydroelectric power ("hydro" for short) was welcomed as an alternative to power from coal-driven and oil-driven generators because it does not consume nonrenewable fossil fuels, and also because it was thought to be environmentally harmless. The latter assumption has proved wrong: the control of rivers by hydro dams has destroyed their natural seasonal variations in flow, disrupting river ecosystems; dams hold back water that would normally flow out to sea in spate at spring runoff, causing the collapse of inshore fisheries; and reservoirs, by drowning large areas of land vegetation, have become a new source of greenhouse gases.[10]

If energy from flowing fresh water has turned out to be a mixed blessing, what about moving seawater? More than 70 percent of the earth's surface is covered by ocean, and it holds vast amounts of energy, as we saw in chapters 6, 7, and 8. The biggest obstacle to harnessing it is its diffuseness: the energy in waves is spread out sideways along the whole width of a wave front. Currents are equally diffuse. And imagine trying to extract useful energy from the rising and falling tide as it creeps up and down the average beach. Many ingenious schemes for capturing this thinly spread energy have been devised, and efforts continue. They are locally successful, on a small scale.

Consider tidal energy first; it differs from other kinds of renewable energy in coming from the rotational energy of the earth-moon-sun system rather than from the sun's radiant energy, which is the source for rivers, waves, and ocean currents. The sites where tidal energy is most concentrated are big estuaries where the tide range is also big. Not many places have these attributes. Two that do are the estuary of the river Rance in northwestern France, which flows into the Gulf of Saint-Malo west of Cherbourg, and the estuary of the Annapolis River in Nova Scotia, which flows into the Bay of Fundy. The maximum range of the tides in the two estuaries are, respectively, 13 m and 17 m (the greatest in the world). Barrages, which function as two-way dams, have been built across the estuary mouths of both rivers to channel the rising and falling tides through narrow gates; the fast-flowing water turns bidirectional turbines that work alternately first in one direction and then the opposite as the tide rises and falls. Not surprisingly, barrages across tidal estuaries affect the local ecology as much as dams on rivers do. A big estuary always flows through low-lying, muddy sediments in a flat coastal plain; interfering with the river's flow inevitably rearranges the mud and adversely affects the water quality and the food available for waterfowl, shorebirds, and fish.[11]

Another way of capturing tidal power is to place underwater turbines where offshore "tide races" produce unusually strong currents. Currents automatically speed up where they are forced to flow through the narrow channels between closely spaced islands, for instance. The number of places where fast offshore tide currents can be exploited greatly exceeds the number of estuaries like those of the Rance and the Annapolis. However, to "block marine channels completely and funnel all the water through . . . turbines," as has been proposed,[12] would affect the marine environment disastrously.

One-way currents, as distinct from two-way tide currents, could also be used to turn turbines, but places where ocean currents are fast enough to make this feasible are not numerous. The Florida Current, whose speed of nearly 8 km/h is exceptionally fast for an ocean current (see chapter 6), is one possibility. Slow-speed turbines would have to be used.

The most obvious manifestations of ocean energy are waves. Their energy can be captured by wave pumps. A wave pump is a hollow cylinder, about 100 m long and open top and bottom, that is held upright in the water so the water level inside it rises and falls as the crests and troughs of the waves go past. The cylinder is heavy enough to hang vertically and is kept afloat in deep water by a buoy anchored to the seabed. On top of the buoy is an air-driven bidirectional turbine. As the water level in the cylinder rises, it forces the air above it up and out at the top; and as the level falls it sucks air in from above. The air, moving first upward and then downward, rotates the turbine and generates an electric current.[13] Electricity generated from waves was first used on navigation buoys to power their warning lamps: the up-and-down movement of a buoy generated electricity to light the lamp it carried.

Using waves to generate large supplies of energy requires a whole array of wave pumps, a veritable wave farm. The pumps are usually aligned in a row, kilometers long, forming an offshore barrage. The barrage is likely to convert the water on its inshore side into an unnaturally calm "lagoon" with reduced water circulation, disturbing the inshore ecosystem.[14]

Waves along a straight shoreline have the drawback of forming a diffuse, spread-out energy source. A wave "power station" sited in a narrow bay can be more compact: the bay funnels the waves and concentrates the energy. Elsewhere, artificial bays—parabolic concrete reflectors to focus the waves—could be used to improve on the natural coastline.[15]

It's worth remarking that none of the methods of harnessing the energy in moving water is altogether environmentally friendly, although it is renewable energy.

*Solar Energy*

To obtain renewable energy from rivers, winds, waves, and currents is to capture solar energy at second hand. In sunny climates, it is simpler to use sunlight as it comes, either as a direct heat source or to generate electricity. Over the past fifty years, more and more people have taken to equipping their homes with solar collectors to provide space heating and hot water. Commonly the sun's heat is absorbed by a flat metal plate collector that transmits the heat to air or water; hot air is stored in a "rock bin," or hot water in an insulated tank, and can be withdrawn as needed.

Alternatively, solar energy can be used to generate electricity by exposing a photoelectric material to sunlight (see chapter 18). This is done in *solar cells,* otherwise known as *photovoltaic cells,* in which a thin wafer of silicon is exposed to the sun. A number of cells are usually wired together in a rectangular array to form a *solar module.*

Solar energy is collected on an industrial scale using parabolic reflectors that track the sun's path across the sky through the daylight hours. This is the most economical way to distill seawater on desert coastlines: the desert conditions give many hours of hot sunshine day after day, providing a dependable energy supply, while at the same time the lack of rain creates a demand for fresh water.

Solar energy trapped by the ocean is also a usable source of energy—in theory. The technology is called OTEC (ocean thermal energy conversion). It uses the strong temperature contrast, in tropical seas, between the warm surface waters and the cold deep waters anywhere from 600 to 1,000 m below the surface. Surface water is about 20°C warmer than water at depth, with negligible changes through the twenty-four hours of day and night.[16] The temperature difference can be used to drive a heat engine, using a volatile fluid such as ammonia as the working fluid. The ammonia is vaporized by the warmth of the surface water and, as a gas, drives a turbine; it then passes to a condensation chamber cooled by deep water from below, where it is converted back into a liquid ready to be vaporized again. It remains to be seen whether the technology will ever become practical on a large scale.

## Geothermal and Nuclear Energy

As we saw in chapter 13, nuclear energy comes from the fission of large nuclei—a process that used to be called "atom splitting"—or by the fusion of small nuclei. Nuclear fission happens naturally on, or rather in, our earth: it goes on all the time in radioactive elements in the earth's crust and mantle, providing the heat that melts rock and turns it into magma, as described in chapter 15. In the neighborhood of volcanoes, where bodies of hot magma collect at no great depth underground, groundwater in the overlying rocks becomes heated; when the hot water emerges at the surface, as hot springs or geysers, it yields the *geothermal energy* that can be harnessed for human use in many parts of the world. People who benefit from it seldom realize that they are, in effect, using energy from a natural nuclear reactor.

Hot springs have been a source of pleasure since time immemorial. They are abundant in Japan, Italy, New Zealand, Iceland, the west coast of North America, Central America, the Philippines, and Indonesia. In some places, especially Iceland, the hot water is piped into buildings to provide free central heating; it heats homes, public buildings, offices, and greenhouses. If the water is boiling, it can supply steam to a geothermal power plant, where it spins the turbines that generate electricity. The hot water remains liquid so long as it is trapped deep in the rocks, under high pressure; it "flashes over" to steam when a "steam well" is drilled down into it, relieving the pressure. A big geothermal power plant must be able to tap copious underground supplies of hot water, trapped under pressure below impermeable "caprock"; then steam can be made to gush at a controlled rate from steam wells drilled through the caprock.[17] At the Geysers, an extremely productive geothermal power site in California that seems to have passed its prime (steam pressures are declining), some of the hundreds of steam wells go down 4,000 m.

From controlling the steam produced by the heat of natural radioactivity, humanity has now progressed to controlling radioactive processes themselves. The first products were "atomic" bombs, or A-bombs; these were soon followed by nuclear reactors used to generate electric power. Reactors are, in effect, "controlled" A-bombs.

The designers of the A-bomb used the fact that when radioactive nuclei split into a pair of smaller nuclei, some neutrons are usually set free because a heavy nucleus contains a higher proportion of neutrons than do the light nuclei formed when it splits. For example, as we noted in chapter 13, when uranium-235 splits spontaneously into one nucleus of strontium-90 and one nucleus of cerium-144, one "leftover" neutron escapes at high speed; only one, because $235 - 90 - 144 = 1$. If the stray neutron strikes another uranium-235 nucleus and embeds itself in it to make highly unstable uranium-236, it imparts additional energy to the newly augmented nucleus, making it split immediately to xenon-140 and strontium-94, plus two leftover neutrons ($236 - 140 - 94 = 2$). The makings of a *chain reaction* have obviously materialized: one naturally produced neutron has led to the formation of two neutrons, which in their turn each produce two more, which each . . . And so on. It happens only if the uranium-235 is sufficiently concentrated for a large proportion of the speeding neutrons to strike other uranium-235 nuclei before being absorbed by comparatively inert material; and once the chain reaction starts, it continues until all the nuclei have been split. If every neutron makes a successful "hit," an exploding A-bomb will yield 20 million joules in one-millionth of a second.[18]

The chain reaction that energizes an A-bomb is slowed down in the nuclear reactors of nuclear power stations, in order to provide a continuous supply of controlled power in place of the uncontrolled explosion of a bomb. The speed of the reaction is controlled by adjusting the richness of the fuel and by dispersing it throughout a *moderator*, a material that absorbs a large proportion of the free neutrons that trigger the successive nuclear fissions. Natural uranium contains only 0.7 percent of strongly radioactive uranium-235; the remaining 99.3 percent is weakly radioactive uranium-238. The uranium-238 functions as a mild moderator by absorbing some of the neutrons emitted by the fissile uranium-235, but in so doing it does not itself split. Materials that perform better as moderators are ordinary water, graphite (pure carbon), and heavy water. Heavy water is water in which ordinary hydrogen, with a single proton as its atomic nucleus, is replaced by deuterium, a hydrogen isotope whose nucleus consists of a proton plus a neutron (see chapter 13); it is a weaker moderator than ordinary water.

All this shows that a reaction rate suitable for a commercial nuclear reactor can be had either by using enriched fuel and a strong moderator or by using natural fuel and a weak moderator. Both strategies are used. Older reactors use the enriched fuel with strong moderator technology. The fuel is enriched with enough added uranium-235 to bring its proportion up to 4 percent, and a strong moderator—ordinary water or graphite—to slow the speeding neutrons and reduce the rate of the chain reaction. The Canadian CANDU reactor uses the other technology. It uses natural uranium—hence less uranium-235—but to keep the reaction going it needs a weaker moderator than ordinary water or graphite; it uses heavy water. The CANDU is the more efficient reactor and also the safer, because its fuel is much less radioactive.

With both kinds of reactor, the heat generated is used to boil ordinary water, to make the steam that spins turbines to generate electricity. That is, ordinary water is the working fluid, irrespective of which kind of water acts as moderator.

A nuclear reactor generating energy at the rate of 1,000 megawatts uses about 3 kilograms of fuel per day.[19] A coal-fired generator of the same power would require about 8,000 tons of fuel per day. But although nuclear reactors supply energy so lavishly, they are not without drawbacks. Their spent fuel is dangerous; it releases injurious radiation and considerable heat for months after it has been discarded, which means it has to be stored in large cooling ponds of (ordinary) water. If things go wrong in a reactor—if the controls malfunction and the reactions speed up too much—a meltdown or an explosion may happen: remember the accidents at Three Mile Island and at Chernobyl, once obscure place-names that are now household words. Even if a reactor behaves perfectly, the waste it produces is dangerously radioactive; disposing of it safely presents an unsolved, and possibly insoluble, problem.

The ideal substitute for fission reactors using radioactive fissionable fuel would be fusion reactors that would mimic energy production in the sun and stars; recall (from chapter 13) that the energy is liberated by the fusion of pairs of heavy hydrogen nuclei to form helium nuclei, releasing tremendous energy as they fuse. Both the unused fuel and the spent fuel are nonradioactive, but the equipment used is left somewhat radioactive. For the reactions—the fusions—to happen, the temperature of the fuel must be raised to millions of degrees Celsius; the need for such extremely high temperatures is the chief difficulty facing engineers trying to devise usable fusion reactors. In a hydrogen bomb (an H-bomb) the necessary high temperature is created by an A-bomb used as a detonator, but this method obviously can't be used to start the fusion reactor in a power plant.

To conclude: at the start of the twenty-first century, the most powerful energy generators are "dirty" nuclear fission reactors. Nuclear fusion reactors would be clean and even more powerful, but at this time (2000), the exploitation of fusion power awaits further advances in nuclear engineering.

Presumably the advances will come. But we must never lose sight of the fact that in nature the "natural reactors" work perfectly: fission energy heats the earth's interior and fusion energy heats the sun. When these two forms of energy are used in man-made nuclear power plants, humanity is doing no more than borrowing a process that operates unceasingly in the natural world.

# EPILOGUE

The time has come to return from the particular to the general. All the details in this book can be boiled down to a few general conclusions.

First, virtually all the *new* energy that keeps our world functioning is nuclear energy, produced by nuclear reactions in which matter is converted to energy according to the famous formula $E = mc^2$.

Second, we also have *old* energy, inherited from the birth of the solar system. Some is residual heat from that event. The rest is the energy that, literally, makes the world go round; it is the energy that causes the earth to rotate and, consequently, causes the tides. This energy is being continuously dissipated by the drag of the tides on the ocean floor. The drag slows the earth's rotation, but so gradually that it will take about 100,000 years for a day to become just one second longer than it is now.

Returning to the new energy: it comes from two sources, the sun and the earth's interior. The solar energy reaching the earth is four thousand times as great as the earth's internal energy: solar energy powers everything we see happening in the sky, the oceans, and the surface of the land, except for volcanoes, hot springs, and earthquakes. It is generated by nuclear fusion in the sun and is transported to us from the sun, across 150 million km, in the form of electromagnetic waves. It is what energizes every living thing on the earth's surface, powering all its actions, its growth, and its reproduction.

The earth's internal energy powers volcanoes, hot springs, and earthquakes and also a variety of unfamiliar living species—bacteria that flourish entirely without the benefit of solar energy. Their habitat is deep in the earth's crust, and also in sulfurous springs—especially the hot springs that emerge through the floor of the deep ocean as hydrothermal vents. This energy is generated by nuclear fission within the body of the earth.

That summarizes what kind of energy we have and where it comes from.

Still to be answered is the question, What *is* energy, anyway? One of the twentieth-century's greatest physicists, Richard Feynman, has called energy an "abstract thing." Others have described it as a "nebulous concept." In any case it cannot be defined. But no more can matter. Everybody knows what *matter* is, but if you try defining it you soon find yourself in a maze of circular arguments. It finally comes down to this: both energy and matter exist, and both are indefinable. They are to physics what axioms are to geometry: just as geometry requires unprovable axioms as a basis for all further deductions, science requires indefinable entities as a jumping-off point for all further discoveries.

Energy and matter can be described as two aspects of the "stuff" the universe is made of. Subatomic particles—the basic units of which every material object is ultimately composed—behave as both energy and matter. This becomes clear when you consider their nature as revealed by modern quantum physics. A moving electron has, simultaneously, the characteristics of both a particle and a train of waves (moving energy). Different observational methods are required to reveal its two aspects. The same is true of radiant energy: it behaves, simultaneously, both as a train of waves and as a stream of moving particles, or "corpuscles."

According to existing knowledge, the stuff of the universe has only two aspects: energy and matter. Perhaps some future scientific breakthrough will reveal other, currently unsuspected aspects as well. It would be arrogant to suppose that scientific discovery will end before *Homo sapiens* goes extinct.

# NOTES

CHAPTER ONE

1. Arthur Beiser, *Concepts of Modern Physics*, 5th ed. (New York: McGraw-Hill, 1995).

CHAPTER TWO

1. Richard P. Feynman, Robert B. Leighton, and Matthew Sands, *The Feynman Lectures on Physics* (Reading, Mass.: Addison-Wesley, 1963), vol. 1, 4-2.

2. John Daintith, *A Dictionary of the Physical Sciences* (London: Macmillan, 1976).

3. This follows from Newton's first law of motion: Every body continues in a state of rest or uniform motion in a straight line until a force acts upon it.

4. Often called free-fall acceleration. For very precise studies a distinction is made between free-fall acceleration, which allows for the slight centrifugal force caused by the earth's rotation on its axis, and pure gravitational acceleration, which would exist on a nonrotating earth. The former is only 0.2 percent less than the latter, and the difference is usually ignored.

5. This is Newton's second law of motion. The unit of force was named the newton in honor of Sir Isaac Newton (1642-1727), the British mathematician and physicist who, among other things, discovered the laws of motion named after him.

6. Named in honor of James Prescott Joule (1818-89), the British scientist who discovered the relationship between heat and mechanical energy.

7. For simplicity, I assume the earth to be a perfect sphere, not flattened at the poles.

8. It is assumed that the plateau is not high enough above sea level for the force of gravity to be appreciably less than at sea level.

9. The reason the Superball does better than this is that in addition to elastic PE, the molecules of the material it is made of have additional chemical PE, which adds to the energy with which it bounces.

10. The distinction between conservative and nonconservative forces given here is incomplete. Further details can be found in a physics textbook such as J. M. Knudsen and P. G. Hjorth, *Elements of Newtonian Mechanics* (New York: Springer, 1995).

CHAPTER THREE

1. Steven Vogel, *Life in Moving Fluids: The Physical Biology of Flow* (Princeton: Princeton University Press, 1981). See also E. C. Pielou, *Fresh Water* (Chicago: University of Chicago Press, 1998).

2. H. J. Gray and Alan Isaacs, *A New Dictionary of Physics*, 2d ed. (London: Longman, 1975).

3. David Halliday, Robert Resnick, and Jearl Walker, *Fundamentals of Physics*, 4th ed. (New York: Wiley, 1993). Note that this refers to *translational* kinetic energy. Molecules consisting of two or more atoms have rotational energy as well.

4. Halliday, Resnick, and Walker, *Fundamentals of Physics*, 582.

5. Named in honor of William Thomson, Lord Kelvin (1824-1907), British physicist.

6. Halliday, Resnick, and Walker, *Fundamentals of Physics.*

7. Paul A. Tipler, *Physics* (New York: Worth, 1976). The equation is a statement of the first law of thermodynamics.

8. An "ideal" engine with the maximum attainable efficiency is called a Carnot engine in honor of Sadi Carnot (1796-1832), the French engineer who discovered the theoretical limit to efficiency.

9. Halliday, Resnick, and Walker, *Fundamentals of Physics*, 607.

10. J. M. Knudsen and P. G. Hjorth, *Elements of Newtonian Mechanics* (New York: Springer, 1995).

11. Knudsen and Hjorth, *Elements of Newtonian Mechanics.*

12. James Thewlis, *Concise Dictionary of Physics and Related Subjects*, 2d ed. (New York: Pergamon Press, 1979).

CHAPTER FOUR

1. Neil Wells, *The Atmosphere and Ocean: A Physical Introduction* (Philadelphia: Taylor and Francis, 1986).

2. The flux of radiation per unit area is often given as approximately $1,360 \text{ W m}^{-2}$; this is the flux per unit area perpendicular to the sun's rays. The flux averaged over the whole spherical surface of the earth is only one-fourth as great.

3. Wells, *Atmosphere and Ocean.* The estimate of fossil fuel reserves relates to the mid-1980s.

4. Arthur Beiser, *Concepts of Modern Physics*, 5th ed. (New York: McGraw-Hill, 1995).

5. For more on solar radiation, see B. Levi, "Are Manufactured Emissions of $CO_2$ Warming Our Climate?" in *Fundamentals of Physics*, 4th ed., by David Halliday, Robert Resnick, and Jearl Walker, 605 (New York: Wiley, 1993).

6. This is known as the earth's *radiative equilibrium temperature.*

7. Figure 4.2 is based on figure 1.6 in Wells, *Atmosphere and Ocean.*

8. C. Donald Ahrens, *Meteorology Today: An Introduction to Weather, Climate, and the Environment*, 4th ed. (St. Paul, Minn.: West, 1991).

9. Paul A. Tipler, *Physics* (New York: Worth, 1976). The effect is named in honor of the French physicist Gustave Gaspard Coriolis (1792-1843).

10. Named in honor of the French physicist Jean Bernard Foucault (1819-68). At the latitude of Paris, 49° N, the plane of the pendulum's swing rotates once in thirty-two hours.

11. Mathematically, the magnitude of the Coriolis acceleration is 2ωs in φ, where ω is the earth's angular velocity of rotation, in radians per second, and φ is the latitude (ω and ? are the Greek letters *omega* and *phi.*

12. Ahrens, *Meteorology Today.*

13. A temporary, local jet stream also develops in the summer over Southeast Asia, India, and Africa, in the zone of east winds just south of the subtropical high; it is known as the *tropical easterly jet stream.*

14. Ahrens, *Meteorology Today.*

15. Air density averages about 1.23 kg m$^{-3}$ at sea level and 0.39 kg m$^{-3}$ at an elevation of 10 km; the kinetic energy of the wind is proportional to $dv^2$, where $d$ is density and $v$ is velocity.

16. Named in honor of Carl Gustav Rossby (1898-1957), the Swedish American meteorologist who discovered the jet streams and developed the theory explaining the meandering route of the polar jet stream.

17. Ahrens, *Meteorology Today.*

CHAPTER FIVE

1. Neil Wells, *The Atmosphere and Ocean: A Physical Introduction* (Philadelphia: Taylor and Francis, 1986).

2. Wells, *Atmosphere and Ocean.*

3. Paul R. Ehrlich, Anne H. Ehrlich, and John P. Holdren, *Ecoscience: Population, Resources, Environment* (San Francisco: Freeman, 1977).

4. Wells, *Atmosphere and Ocean.*

5. For a mathematical treatment, see Stephen Pond and George L. Pickard, *Introductory Dynamical Oceanography*, 2d ed. (New York: Pergamon Press, 1983).

6. C. Donald Ahrens, *Meteorology Today*, 4th ed. (St. Paul, Minn.: West, 1991).

7. Ahrens, *Meteorology Today.*

8. David Brunt, *Weather Study* (London: Thomas Nelson, 1942), 136.

9. E. C. Pielou, *Fresh Water* (Chicago: University of Chicago Press, 1998).

10. Wells, *Atmosphere and Ocean*.

11. Ahrens, *Meteorology Today*.

12. Wells, *Atmosphere and Ocean*.

13. These numbers are derived from figure 9.4 in Wells, *Atmosphere and Ocean*.

14. M. Grant Gross and Elizabeth Gross, *Oceanography: A View of Earth*, 7th ed. (Upper Saddle River, N.J.: Prentice Hall, 1996).

15. Wells, *Atmosphere and Ocean*.

16. Ahrens, *Meteorology Today*.

17. The velocity, $v$, is found from the equation $v^2 = 2gx$, where $g = 9.81$ m s$^{-2}$ is the acceleration due to gravity and $x = 200$ m is the distance fallen.

CHAPTER SIX

1. Neil Wells, *The Atmosphere and Ocean: A Physical Introduction* (Philadelphia: Taylor and Francis, 1985). See also C. Donald Ahrens, *Meteorology Today*, 4th ed. (St. Paul, Minn.: West, 1991).

2. Wells, *Atmosphere and Ocean*.

3. George L. Pickard and William J. Emery, *Descriptive Physical Oceanography: An Introduction*, 5th ed. (New York: Pergamon Press, 1990).

4. Pickard and Emery, *Descriptive Physical Oceanography*.

5. Wells, *Atmosphere and Ocean*.

6. Wells, *Atmosphere and Ocean*. A rise of temperature as great as this would happen (theoretically) at a latitude of 60°, where a summer day can be eighteen hours long.

7. Richard E. Thomson, *Oceanography of the British Columbia Coast* (Ottawa: Department of Fisheries and Oceans, 1981).

8. Steven Vogel, *Life in Moving Fluids: The Physical Biology of Flow* (Princeton: Princeton University Press, 1981), 69.

9. Strictly, *dynamic viscosity*, the drag between the very thin layers that slide over each other in a moving fluid.

10. Vogel, *Life in Moving Fluids*, table 2.1.

11. See Dale E. Ingmanson and William J. Wallace, *Oceanography: An Introduction*, 4th ed. (Belmont, Calif.: Wadsworth, 1989), for the speeds of the Florida Current and the Gulf Stream off Europe.

12. The first two winds in the table are of force 10 and force 2, respectively, on the Beaufort scale of wind force.

13. Vogel, *Life in Moving Fluids*.

14. Stephen Pond and George L. Pickard, *Introductory Dynamical Oceanography*, 2d ed. (New York: Pergamon Press, 1983).

15. Vagn Walfrid Ekman (1874-1954), Swedish oceanographer and physicist.

16. Pond and Pickard, *Introductory Dynamical Oceanography*.

17. Thomson, *Oceanography of the British Columbia Coast*.

18. This is an Ekman spiral in the atmosphere. The name is used less in meteorology than in oceanography.

19. Pond and Pickard, *Introductory Dynamical Oceanography*.

20. Pond and Pickard, *Introductory Dynamical Oceanography.*

21. M. Grant Gross and Elizabeth Gross, *Oceanography: A View of Earth,* 7th ed. (Upper Saddle River, N.J.: Prentice Hall, 1996).

22. Pond and Pickard, *Introductory Dynamical Oceanography.* The drop in pressure amounts to about 3 kilopascals = 30 millibars.

23. Gross and Gross, *Oceanography.*

24. C. Donald Ahrens, *Meteorology Today,* 4th ed. (St. Paul, Minn.: West, 1991).

25. A pycnocline is described as a *thermocline* if a fall in temperature is the predominant change or a *halocline* if a drop in salinity predominates.

26. Gross and Gross, *Oceanography.*

27. See Ingmanson and Wallace, *Oceanography.*

28. The map is based on W. S. Broecker, "Thermohaline Circulation, the Achilles Heel of Our Climate System: Will Man-Made $CO_2$ Upset the Current Balance?" *Science* 278 (1997): 1582-88; W. S. Broecker and G. H. Denton, "The Great Ocean Conveyor," *Oceanography* 4 (1991): 79-89; Gross and Gross, *Oceanography;* and Arthur L. Bloom, *Geomorphology: A Systematic Analysis of Late Cenozoic Landforms,* 3d ed. (Upper Saddle River, N.J.: Prentice Hall, 1998).

29. Broecker, "Thermohaline Circulation."

30. Broecker and Denton, "Great Ocean Conveyor," 54.

31. Wells, *Atmosphere and Ocean.*

CHAPTER SEVEN

1. Stephen Pond and George L. Pickard, *Introductory Dynamical Oceanography,* 2d ed. (New York: Pergamon Press, 1983).

2. See Dale E. Ingmanson and William J. Wallace, *Oceanography: An Introduction,* 4th ed. (Belmont, Calif.: Wadsworth, 1989), 208.

3. Richard E. Thomson, *Oceanography of the British Columbia Coast* (Ottawa: Department of Fisheries and Oceans, 1981), 106.

4. Pond and Pickard, *Introductory Dynamical Oceanography,* 234, table 12.4. The wave heights given in the table are "significant wave heights," $H_s$. Average wave height $= 0.6 \times H_s$.

5. Not in exactly the same place: the log would drift slowly downwind, at 3/1,000 of the wind speed, because of *Stokes drift.*

6. Thomson, *Oceanography of the British Columbia Coast.*

7. Pond and Pickard, *Introductory Dynamical Oceanography.*

8. Pond and Pickard, *Introductory Dynamical Oceanography.*

9. Mathematically, the analysis is into a Fourier series.

10. Wave speed, $C$, and wavelength, $L$, are related by the formula $C = \sqrt{(Lg/2\pi)}$; $g$ is the acceleration due to gravity. The formula applies only to gravity waves in deep water.

11. Thomson, *Oceanography of the British Columbia Coast.*

12. Thomson, *Oceanography of the British Columbia Coast.*

13. Pond and Pickard, *Introductory Dynamical Oceanography.*

14. Pond and Pickard, *Introductory Dynamical Oceanography.*

15. Thomson, *Oceanography of the British Columbia Coast.*

16. Thomson, *Oceanography of the British Columbia Coast.*

17. Edward J. Tarbuck and Frederick K. Lutgens, *Earth: An Introduction to Physical Geology,* 4th ed. (New York: Macmillan, 1993).

18. The energy in a tsunami is inversely proportional to the distance from its source. Note the contrast with sound waves, whose energy is inversely proportional to the square of the distance. This is because water waves travel across a two-dimensional surface, sound waves through a three-dimensional volume. A ring of water waves has length $2\pi r$; a spherical shell of sound waves has area $4\pi r^2$.

19. *Waves, Tides, and Shallow Water Processes* (New York: Pergamon in association with Open University, 1989).

20. The discovery is described on the Web page of Heriot-Watt University, Edinburgh, (2000). Quotations from Russell's writings given here are excerpted from this source.

21. R. Herman, "Solitary Waves," *American Scientist* 80 (1992): 350-61. The article gives a nontechnical account of the physics of solitary waves and contains a satellite photo of them in the Andaman Sea.

22. M. K. Ramamurthy, B. P. Collins, R. M. Rauber, and P. C. Kennedy, "Evidence of Very-Large-Amplitude Solitary Waves in the Atmosphere," *Nature* 348 (1990): 314-17.

23. For more on solitons, see Paul Davies and John Gribbin, *The Matter Myth* (New York: Simon and Schuster, 1992).

CHAPTER EIGHT

1. Often called "tidal" currents and "tidal" streams. The adjective "tide" is used here to correspond with "tide wave."

2. Richard E. Thomson, *Oceanography of the British Columbia Coast* (Ottawa: Department of Fisheries and Oceans, 1981).

3. The moon's orbit is slightly elliptical; the distance from earth to moon varies between a minimum of 363,000 km and a maximum of 406,000 km.

4. Strictly (and pedantically) speaking, "centrifugal force" is a *pseudoforce,* that is, not a true force but simply a reaction to centripetal force.

5. Edward J. Tarbuck and Frederick K. Lutgens, *Earth: An Introduction to Physical Geology,* 4th ed. (New York: Macmillan, 1993).

6. Stephen Pond and George L. Pickard, *Introductory Dynamical Oceanography,* 2d ed. (New York: Pergamon Press, 1983).

7. B. F. Chao and R. S. Gross, "Changes in the Earth's Rotational Energy Induced by Earthquakes," *Geophysical Journal International* 122 (1995): 776-83.

8. Pond and Pickard, *Introductory Dynamical Oceanography.*

CHAPTER NINE

1. Edward J. Tarbuck and Frederick K. Lutgens, *Earth: An Introduction to Physical Geology,* 4th ed. (New York: Macmillan, 1993).

2. This also happens in metamorphic rocks.

3. Arthur L. Bloom, *Geomorphology: A Systematic Analysis of Late Cenozoic Landforms*, 3d ed. (Upper Saddle River, N.J.: Prentice Hall, 1998).

4. W. A. E. McBryde and R. P. Graham, *The Outlines of Chemistry* (Toronto: Clarke, Irwin, 1966).

5. J. E. Gordon, *The New Science of Strong Materials, or Why You Don't Fall through the Floor*, 2d ed. (New York: Viking Penguin, 1984).

6. Gordon, *New Science of Strong Materials*.

7. Luna B. Leopold, *A View of the River* (Cambridge: Harvard University Press, 1994).

8. Chester B. Beaty, *The Landscapes of Southern Alberta* (Lethbridge, Alta.: University of Lethbridge, 1975).

9. H. J. Melosh, "Acoustic Fluidization," *American Scientist* 71 (1983): 158-65. See also Beaty, *Landscapes of Southern Alberta*.

10. Melosh, "Acoustic Fluidization."

11. Melosh, "Acoustic Fluidization," 159.

12. After the ice sheets of the last ice age melted, while the climate was dry, the winds were strong, and the land was unvegetated, wind erosion was more important than river erosion. See E. C. Pielou, *After the Ice Age* (Chicago: University of Chicago Press, 1991).

13. Particles of clay have diameters in the range 0.0005 to 0.0039 mm; silt, 0.0039 to 0.0625 mm; and sand, 0.0625 to 2.0 mm. See Nancy D. Gordon, T. A. McMahon, and Brian L. Finlayson, *Stream Hydrology* (New York: Wiley, 1992).

14. Leopold, *View of the River*.

15. For details, see Gordon, McMahon, and Finlayson, *Stream Hydrology*.

16. C. H. Crickmay, *The Work of the River* (London: Macmillan, 1974). Note that these are terrestrial surfaces. Submarine erosion shifts ten times as much material on the seafloor.

17. M. G. Wolman, "The Impact of Man," *Eos* 71 (1990): 1884-86.

CHAPTER TEN

1. William L. Masterton and Emil J. Slowinski, *Chemical Principles* (Philadelphia: Saunders, 1977).

2. There are also *metallic bonds,* which hold metals together when they occur in pure form, uncombined with other elements. Pure metals are rare in nature; exceptions are gold and occasionally copper.

3. Masterton and Slowinski, *Chemical Principles*.

4. J. E. Gordon, *The New Science of Strong Materials, or Why You Don't Fall through the Floor*, 2d ed. (New York: Viking Penguin, 1984).

5. Note that it is not possible to calculate the energy change in a reaction directly from the bond energies of the reactants unless all the reactants, and the product, are gases.

6. Free energy has two other names. Chemists, who normally carry out experiments at constant pressure, write of *Gibbs energy.* Physicists experimenting with gases, and keeping volume constant while allowing pressure to vary, write of *Helmholtz energy.*

The two are the same if pressure makes no difference to the outcome; see Michael M. Abbott and H. C. Van Ness, *Thermodynamics,* 2d ed. (New York: McGraw-Hill, 1989).

7. H. Robert Horton, L. A. Moran, R. S. Ochs, J. D. Rawn, and K. G. Scrimgeour, *Principles of Biochemistry* (Upper Saddle River, N.J.: Prentice Hall, 1993).

8. David Halliday, Robert Resnick, and Jearl Walker, *Fundamentals of Physics,* 4th ed. (New York: Wiley, 1993).

9. Masterton and Slowinski, *Chemical Principles.*

10. Horton et al., *Principles of Biochemistry.*

CHAPTER ELEVEN

1. Wavelengths are often measured in nanometers (abbreviation nm). 1 nm = $1 \times 10^{-3}$ $\mu$m = $1 \times 10^{-9}$ m. An older unit, now seldom used, is the *angstrom,* with a length of $10^{-10}$ m (0.1 nm).

2. H. Robert Horton, L. A. Moran, R. S. Ochs, J. D. Rawn, and K. G. Scrimgeour, *Principles of Biochemistry* (Upper Saddle River, N.J.: Prentice Hall, 1993).

3. David H. Miller, *Energy at the Surface of the Earth: An Introduction to the Energetics of Ecosystems* (New York: Academic Press, 1981).

4. Paul R. Ehrlich, Anne H. Ehrlich, and John P. Holdren, *Ecoscience: Population, Resources, Environment* (San Francisco: Freeman, 1977).

5. N. L. Stephenson, "Climate Control of Vegetation Distribution: The Role of Water Balance," *American Naturalist* 135 (1990): 649-70. See also E. C. Pielou, *Fresh Water* (Chicago: University of Chicago Press, 1998).

6. F. B. Golley, "Energy Flux in Ecosystems," in *Ecosystem Structure and Function,* ed. J. A. Wiens, 60-90, Proceedings of the Thirty-first Annual Biology Colloquium (Corvallis: Oregon State University, 1972).

7. This is the average of two estimates, one of $1.6 \times 10^{18}$ kJ given by Golley, "Energy Flux in Ecosystems," and one of $2.0 \times 10^{18}$ kJ given by Ehrlich, Ehrlich, and Holdren, *Ecoscience.*

8. Ehrlich, Ehrlich, and Holdren, *Ecoscience.*

9. G. M. Woodwell, "Aquatic Systems as Part of the Biosphere," in *Fundamentals of Aquatic Ecosystems,* ed. R. S. K. Barnes and K. H. Mann, 201-15 (Boston: Blackwell, 1980).

10. Ehrlich, Ehrlich, and Holdren, *Ecoscience.*

11. Ehrlich, Ehrlich, and Holdren, *Ecoscience.* See also Woodwell, "Aquatic Systems as Part of the Biosphere."

12. The rate is known as *irradiance* and is measure in joules per square meter (J m$^{-2}$).

13. G. E. Fogg, "Phytoplankton Primary Production," in *Fundamentals of Aquatic Ecosystems,* ed. R. S. K. Barnes and K. H. Mann, 24-45 (Boston: Blackwell, 1980).

14. Fogg, "Phytoplankton Primary Production." See also Ehrlich, Ehrlich, and Holdren, *Ecoscience.*

15. R. S. K. Barnes, "The Unity and Diversity of Aquatic Systems," in *Fundamentals of Aquatic Ecosystems,* ed. R. S. K. Barnes and K. H. Mann, 5-23 (Boston: Blackwell, 1980).

16. Ehrlich, Ehrich, and Holdren, *Ecoscience.*

17. William L. Masterton and Emil J. Slowinski, *Chemical Principles,* 4th ed. (Philadelphia: Saunders, 1977).

CHAPTER TWELVE

1. This formula is a simplification; more precisely, the formula for carbohydrates is $(CH_2O)_n$, with $n$ not less than 3.

2. L. R. Pomeroy, "Detritus and Its Role as a Food Source," in *Fundamentals of Aquatic Ecosystems,* ed. R. S. K. Barnes and K. H. Mann, 84-102 (Boston: Blackwell, 1980).

3. Pomeroy, "Detritus and Its Role as a Food Source," 90.

4. J. F. Franklin and J. M. Hemstrom, "Aspects of Succession in Coniferous Forests of the Pacific Northwest," in *Forest Succession: Concepts and Application,* ed. Darrell C. West, Herman H. Shugart, and Daniel B. Botkin, 212-29 (New York: Springer-Verlag, 1981).

5. Alternative terms are *chemoautotrophy* and *chemolithotrophy.*

6. Sergei Nikolaevich Vinogradsky (or Winogradsky) (1856-1953).

7. Quoted in Michael T. Madigan, John M. Martinko, and Jack Parker, *Brock Biology of Microorganisms,* 7th ed. (Upper Saddle River, N.J.: Prentice Hall, 1994).

8. E. C. Pielou, *Fresh Water* (Chicago: University of Chicago Press, 1998).

9. The bacteria concerned are *Nitrosomonas* and *Nitrobacter.* Most of the ammonia is made by the nitrogen-fixing bacteria *Rhizobium,* which live in the nodules on the roots of leguminous plants.

10. J. P. Barry and R. E. Kochevar, "*Calyptogena diagonalis,* a New Vesicomyid Bivalve from Subduction Zone Cold Seeps in the Eastern North Pacific," *Veliger* 42 (1999): 117-23. See also A. Ahmad, J. P. Barry, and D. C. Nelson, "Phylogenetic Affinity of a Wide, Vacuolate, Nitrate-Accumulating *Beggiatoa* sp. from Monterey Canyon, California, with *Thioploca* spp.," *Applied and Environmental Microbiology* 65 (1999): 270-77.

11. R. N. Nikolaus, "*Beggiatoa* and Hydrocarbon Seeps," *Quarterdeck* 5 (1997): 2.

12. Edward J. Tarbuck and Frederick J. Lutgens, *Earth: An Introduction to Physical Geology,* 4th ed. (New York: Macmillan, 1993).

13. P. Westbroek, *Life as a Geological Force: Dynamics of the Earth* (New York: Norton, 1992).

14. See Westbroek, *Life as a Geological Force,* for an illustrated account of the details.

15. Westbroek, *Life as a Geological Force,* 139.

16. Ralph Buchsbaum and Lorus J. Milne, *The Lower Animals: Living Invertebrates of the World* (Garden City, N.Y.: Doubleday, 1966), 31.

CHAPTER THIRTEEN

1. C. M. R. Fowler, *The Solid Earth: An Introduction to Global Geophysics* (Cambridge: Cambridge University Press, 1990).

2. Arthur Beiser, *Concepts of Modern Physics,* 5th ed. (New York: McGraw-Hill, 1995).

3. Beiser, *Concepts of Modern Physics.*

4. Protons are somewhat lighter than neutrons; their masses are, respectively, $1.673 \times 10^{-30}$ g and $1.675 \times 10^{-30}$ g.

5. William L. Masterton and Emil J. Slowinski, *Chemical Principles* (Philadelphia: Saunders, 1977).

6. Masterton and Slowinski, *Chemical Principles*.

7. Beiser, *Concepts of Modern Physics*.

8. Masterton and Slowinski, *Chemical Principles*.

9. Fowler, *Solid Earth*.

CHAPTER FOURTEEN

1. M. Wysession, "The Inner Workings of the Earth," *Scientific American*, March–April 1995.

2. C. M. R. Fowler, *The Solid Earth: An Introduction to Global Geophysics* (Cambridge: Cambridge University Press, 1990).

3. Fowler, *Solid Earth*.

4. See, for example, Edward J. Tarbuck and Frederick K. Lutgens, *Earth: An Introduction to Physical Geology*, 4th ed. (New York: Macmillan, 1993).

5. Fowler, *Solid Earth*.

6. Wysession, "Inner Workings of the Earth."

CHAPTER FIFTEEN

1. R. D. van der Hilst, S. Widiyantoro, and E. R. Engdahl, "Evidence for Deep Mantle Circulation from Global Tomography," *Nature* 386 (1997): 578-84. See also J. A. Jacobs, *The Deep Interior of the Earth* (New York: Chapman and Hall, 1992).

2. C. M. R. Fowler, *The Solid Earth: An Introduction to Global Geophysics* (Cambridge: Cambridge University Press, 1990).

3. M. Wysession, "The Inner Workings of the Earth," *Scientific American*, March–April 1995.

4. Fowler, *Solid Earth*.

5. R. van der Voo, W. Spakman, and H. Bijwaard, "Mesozoic Subducted Slabs under Siberia," *Nature* 397 (1999): 246-49. See also M. A. Richards, "Prospecting for Jurassic Slabs," *Nature* 397 (1999): 203-4.

6. B. F. Chao, R. S. Gross, and D.-N. Dong, "Changes in Global Gravitational Energy Induced by Earthquakes," *Geophysical Journal International* 122 (1995): 784-89.

7. Fowler, *Solid Earth*, 234.

8. Wysession, "Inner Workings of the Earth."

9. M. Grant Gross and Elizabeth Gross, *Oceanography: A View of the Earth*, 7th ed. (Upper Saddle River, N.J.: Prentice Hall, 1996). See also Fowler, *Solid Earth*.

10. Gross and Gross, *Oceanography*.

11. Fowler, *Solid Earth*, 301.

12. V. Tunnicliffe and C. M. R. Fowler, "Influence of Sea-Floor Spreading on the Global Hydrothermal Vent Fauna," *Nature* 379 (1996): 531-33.

13. R. Flanagan, "The Light at the Bottom of the Sea," *New Scientist* 156 (1997): 42-46.

14. B. F. Chao and R. S. Gross, "Changes in the Earth's Rotational Energy Induced by Earthquakes," *Geophysical Journal International* 122 (1995): 776-83.

15. Chao, Gross, and Dong, "Changes in Global Gravitational Energy Induced by Earthquakes."

16. Fowler, *Solid Earth*. See also Richard V. Fisher, Grant Heiken, and Jeffrey B. Hullen, *Volcanoes: Crucibles of Change* (Princeton: Princeton University Press, 1997), and Edward J. Tarbuck and Frederick K. Lutgens, *The Earth: An Introduction to Physical Geology*, 4th ed. (New York: Macmillan, 1993).

17. Wysession, "Inner Workings of the Earth."

18. Tarbuck and Lutgens, *Earth*; see also Shawna Vogel, *Naked Earth: The New Geophysics* (New York: Dutton, 1995).

19. At present the Pacific plate underlies most of the ocean floor; it has replaced earlier plates that have now disappeared under the North American continent.

20. Ben Gadd, *Handbook of the Canadian Rockies* (Jasper, Alta.: Corax Press, 1986).

21. Chester R. Longwell, Richard Foster Flint, and John E. Sanders, *Physical Geology* (New York: Wiley, 1969).

22. John Verhoogen, *Energetics of the Earth* (Washington, D.C.: National Academy of Sciences, 1980).

23. This is *plastic deformation*, in contrast to impermanent *elastic deformation*.

CHAPTER SIXTEEN

1. See, for example, John R. Gribbin and Mary Gribbin, *Richard Feynman: A Life in Science* (New York: Dutton, 1997).

2. See, for example, Sybil P. Parker, *McGraw-Hill Dictionary of Physics*, 2d ed. (New York: McGraw-Hill, 1997), or H. J. Gray and Alan Isaacs, *A New Dictionary of Physics*, 2d ed. (London: Longman, 1975).

3. This is not to say that all conductors are equally efficient, or all insulators, and it disregards semiconductors. For details see, for example, Arthur Beiser, *Concepts of Modern Physics*, 5th ed. (New York: McGraw-Hill, 1995).

4. J. E. Gordon, *The New Science of Strong Materials, or Why You Don't Fall through the Floor*, 2d ed. (New York: Viking Penguin, 1984), 271.

5. Gray and Isaacs, *New Dictionary of Physics*.

6. Voltage is otherwise known as *potential difference* and is measured in volts.

7. NASA, (1999).

8. See any physics textbook; for example, David Halliday, Robert Resnick, and Jearl Walker, *Fundamentals of Physics*, 4th ed. (New York: Wiley, 1993).

9. R. A. Black and J. Hallett, "The Mystery of Cloud Electrification," *American Scientist*, November–December, 1998.

10. See, for example, C. Donald Ahrens, *Meteorology Today*, 4th ed. (St. Paul, Minn.: West, 1991).

11. The end of the compass needle that points northward, toward the earth's north magnetic pole (latitude 78° N, longitude 104° W), has become known as the needle's north pole. This is a tradition unlikely to change, and it fixes the way the poles of all

magnets are labeled. Consequently, the north magnetic pole is equivalent to a south pole in the physicist's sense; correspondingly, the south magnetic pole is equivalent to a north pole.

12. Halliday, Resnick, and Walker, *Fundamentals of Physics*.

13. Magnetite is the iron oxide $Fe_3O_4$; it is a black mineral with metallic luster.

14. Roger G. Newton, *What Makes Nature Tick?* (Cambridge: Harvard University Press, 1993).

15. Other forms of magnetism include *paramagnetism* and *diamagnetism* (see a textbook of physics for definitions). They are of comparatively minor importance in the context of the earth's energy budget, so they are not discussed in this book.

16. P. C. W. Davies, *The Forces of Nature* (New York: Cambridge University Press, 1979), 19. This book summarizes the historical development of the subject as well as explaining modern concepts.

17. André-Marie Ampère (1775-1836) and Charles-Augustin Coulomb (1736-1806).

18. C. M. R. Fowler, *The Solid Earth: An Introduction to Global Geophysics* (New York: Cambridge University Press, 1990).

CHAPTER SEVENTEEN

1. This book discusses traveling or progressive waves, here simply called "waves." Standing or stationary waves, which stay in one place, are described in most physics textbooks.

2. Respectively, John Daintith, *A Dictionary of the Physical Sciences* (London: Macmillan, 1976), 320, and Sybil P. Parker, *McGraw-Hill Dictionary of Physics*, 2d ed. (New York: McGraw-Hill, 1997), 470.

3. David Halliday, Robert Resnick, and Jearl Walker, *Fundamentals of Physics*, 4th ed. (New York: Wiley, 1993), 476.

4. Quasi-stellar radio sources, possibly the farthest (from us) and most powerful objects in the universe.

5. Wave energy varies over the time it takes for a wave to pass a given point. Its energy is measured as an *average* over the whole wave.

6. Douglas C. Giancoli, *Physics*, 3d ed. (Englewood Cliffs, N.J.: Prentice Hall, 1991), 311.

7. The original unit was the *bel*, named in honor of Alexander Graham Bell (1847-1922). It is an inconveniently large unit, so the decibel (0.1 bel) is the unit most often used.

8. Another unit of sound intensity is the *phonon*. It is a subjective measure, designed to allow for the fact that sounds of the same dBs but at different frequencies don't seem equally loud to the hearer.

9. These speeds assume the temperature of the medium to be 20°C. As the temperature falls, the speed increases.

10. The speed is $v = \sqrt{(B/\rho)}$ m s$^{-1}$, where $B$ denotes the bulk modulus of the medium and $\rho$ is its density ($\rho$ is the Greek letter *rho*).

11. Two other kinds of surface (or near surface) waves are *Love waves,* in which particle motion is horizontal, and *Stonely waves,* which follow the surfaces between rock strata.

12. The unit for frequency is the hertz, abbreviated Hz (named for the German physicist Heinrich Rudolf Hertz, 1857-94). One hertz equals one wave per second. As an alternative to frequency (*f*), period $T = 1/f$ seconds is often used. Thus, average periods for P-, S-, and R-waves are 20, 17, and 25 seconds, respectively.

13. C. M. R. Fowler, *The Solid Earth: An Introduction to Global Geophysics* (Cambridge: Cambridge University Press, 1990).

14. All seismic waves can propagate through a compressible medium. S-waves and R-waves can propagate only if the medium has *rigidity* as well as *compressibility.*

15. Charles Francis Richter (1900-1985).

16. British Geological Survey, (1999).

17. This assumes that the energy, *E,* has been computed from the surface wave magnitude, $M_s$, using Bath's equation, $\log_{10}E = 5.24 + 1.44M_s$. See Fowler, *Solid Earth,* 93.

18. This may have been a seismically quiet period; seismic activity was perhaps an order of magnitude greater in a fifteen-year period studied earlier (1950-65). See B. F. Chao, R. S. Gross, and D.-N. Dong, "Changes in Global Gravitational Energy Induced by Earthquakes," *Geophysical Journal International* 122 (1995): 784-89.

19. B. F. Chao and R. S. Gross, "Changes in the Earth's Rotational Energy Induced by Earthquakes," *Geophysical Journal International* 122 (1995): 776-83; see also Chao, Gross, and Dong, "Changes in Global Gravitational Energy Induced by Earthquakes."

CHAPTER EIGHTEEN

1. In technical terms, *E* is the electric field intensity and *B* is the magnetic flux density.

2. The equation usually appears as $E = hf$, where *f* is the frequency; $E = hf$ is the appropriate formula when the energy is propagating not through a vacuum but through a medium, such as water, in which the velocity is less than *c.*

3. Arthur Beiser, *Concepts of Modern Physics,* 5th ed. (New York: McGraw-Hill, 1995). Braking radiation is usually known by its German name, *Bremsstrahlung.*

4. Virgilio Acosta, Clyde L. Cowan, and B. J. Graham, *Essentials of Modern Physics* (New York: Harper and Row, 1973).

5. Acosta, Cowan, and Graham, *Essentials of Modern Physics.*

6. Douglas C. Giancoli, *Physics,* 3d ed. (Englewood Cliffs, N.J.: Prentice Hall, 1991).

7. P. C. W. Davies, *The Forces of Nature* (New York: Cambridge University Press, 1979), 23.

8. They do this in two ways: by *pair production,* caused by γ-rays, and by Compton scattering, caused by X rays. See, for example, Beiser, *Concepts of Modern Physics.*

9. E. C. Pielou, *The World of Northern Evergreens* (Ithaca: Cornell University Press, 1988).

10. William L. Masterton and Emil J. Slowinski, *Chemical Principles,* 4th ed. (Philadelphia: Saunders, 1977).

11. Beiser, *Concepts of Modern Physics;* see also Acosta, Cowan, and Graham, *Essentials of Modern Physics.*

12. Paul A. Tipler, *Physics* (New York: Worth, 1976), 627.

13. The molecules' diameters are about 0.2 nm ($2 \times 10^{-9}$ m); the longest wavelength of visible light is about 0.4 μm ($4 \times 10^{-7}$ m).

14. H. J. Gray and Alan Isaacs, *A New Dictionary of Physics,* 2d ed. (London: Longman, 1975).

CHAPTER NINETEEN

1. Thomas Gold, *The Deep Hot Biosphere* (New York: Copernicus, 1999).

2. Jared Diamond, *The Third Chimpanzee: The Evolution and Future of the Human Animal* (New York: HarperCollins, 1992). See also I. Tattersall, "Once We Were Not Alone," *Scientific American* 282 (January 2000): 57-67, for the dates.

3. Brian J. Skinner and Stephen C. Porter, *The Dynamic Earth: An Introduction to Physical Geology,* 3d ed. (New York: Wiley, 1995).

4. Skinner and Porter, *Dynamic Earth.*

5. Gold, *Deep Hot Biosphere.*

6. Search the Internet for Solar Hydrogen Energy Corporation.

7. Search the Internet for Ballard Fuel Cell.

8. E. Suess, G. Bohrmann, J. Greinert, and E. Lausch, "Flammable Ice," *Scientific American* 281 (November 1999): 76-83.

9. Marc Reisneer, *Cadillac Desert: The American West and Its Disappearing Water,* rev. ed. (New York: Penguin Books, 1993).

10. E. C. Pielou, *Fresh Water* (Chicago: University of Chicago Press, 1998).

11. Harry Thurston, *Tidal Life: A Natural History of the Bay of Fundy* (Willowdale, Ont.: Fireside Books, 1990).

12. F. Pearce, "Catching the Tide," *New Scientist* 20 (June 1998): 38-41.

13. Willard Bascom, *Waves and Beaches: The Dynamics of the Ocean Surface,* rev. ed. (Garden City, N.Y.: Anchor Press, 1980).

14. *Waves, Tides, and Shallow Water Processes* (London: Pergamon in association with Open University, 1989).

15. B. Clayton, "Bay Wash," *New Scientist,* 1 November 1997, 6.

16. Dale E. Ingmanson and William J. Wallace, *Oceanography: An Introduction,* 4th ed. (Belmont, Calif.: Wadsworth, 1989).

17. Richard V. Fisher, Grant Heiken, and Jeffrey B. Hullen, *Volcanoes: Crucibles of Change* (Princeton: Princeton University Press, 1997).

18. Arthur Beiser, *Concepts of Modern Physics,* 5th ed. (New York: McGraw-Hill, 1995).

19. F. J. Shore, "Nuclear Reactors," in *Encyclopedia of Physics,* 2d ed., ed. Rita G. Lerner and George L. Trigg (New York: VCH, 1991).

# INDEX

Pielou, E. C.,
1924-

The energy of
nature.

| DATE | | |
|------|------|------|
|  |  |  |
|  |  |  |
|  |  |  |
|  |  |  |
|  |  |  |
|  |  |  |
|  |  |  |
|  |  |  |
|  |  |  |
|  |  |  |
|  |  |  |
|  |  |  |
|  |  |  |
|  |  |  |